HANDS-ON ASTRO

HANDS-ON ASTRONOMY

A Cambridge Guide to Equipment and Accessories

HERVÉ BURILLIER
AND
CHRISTOPHE LEHÉNAFF

TRANSLATED BY
KLAUS BRASCH

WITH UPDATES BY
MICHAEL COVINGTON

CAMBRIDGE
UNIVERSITY PRESS

PUBLISHED BY THE PRESS SYNDICATE OF THE UNIVERSITY OF CAMBRIDGE
The Pitt Building, Trumpington Street, Cambridge, United Kingdom

CAMBRIDGE UNIVERSITY PRESS
The Edinburgh Building, Cambridge CB2 2RU, UK
40 West 20th Street, New York, NY 10011–4211, USA
477 Williamstown Road, Port Melbourne, VIC 3207, Australia
Ruiz de Alarcón 13, 28014 Madrid, Spain
Dock House, The Waterfront, Cape Town 8001, South Africa

http://www.cambridge.org

First published by BORDAS in 1998
GUIDE DU MATERIEL D'ASTRONOMIE
© BORDAS/VUEF
Second edition 2001

Printed in Italy by Canale & C.Sp.A.Torino

ISBN 0 521 00598 1 Paperback

The publisher thanks the following companies for the use of their photographs: La
Maison de l'Astronomie (Paris), Meade Instruments (Irvine, Carlifornia), Médas (Vichy,
France), Optique Unterlinden (Colmar, France), and S.P.J.P. Paralux (Créteil, France).

The authors thank Sandrine Léglise, Annie and Bernard Olveira, Micheline and Bernard
Lehénaff, Patrick Pelletier, and Didier Avoscan.

Contents

Preface

Amateur astronomy is a most engaging pastime which is constantly attracting new enthusiasts. Whether you enjoy observing lunar craters or planetary details, admire great vistas of thousands of stars, or witness almost first-hand the birth of new suns in distant nebulae, recreational astronomy is a highly enriching personal experience.

Whether novice or highly experienced, amateur astronomers today are faced with an abundant and diverse selection of commercial equipment and accessories. The authors, who deal professionally in the sale and utilization of astronomical equipment and accessories, have compiled as much practical and basic information as possible in this book, to assist readers in selecting binoculars, telescopes and other materials most suitable to them.

Even after you have decided on what equipment you want and are ready to make a purchase, be sure to consult others, get informed and shop around; while astronomy is a rapidly growing pastime it is unlikely that the first store you visit will meet all your needs. Also keep

The authors' choice

The equipment described in this volume was selected by the authors as most representative of current trends and developments in amateur astronomy. All performance and quality specifications provided by manufacturers were verified first-hand through equipment and material field tests, and price quotes reflect those advertised in popular astronomy magazines and periodicals. Our recommendations have taken into consideration not only the needs of the amateur, but also such factors as location, budget, level of knowledge and degree of interest. Lastly, we have deliberately omitted any mention of amateur telescope making. This has become largely the domain of individuals who enjoy telescope building for its own sake, rather than, as in the past, an economic necessity for most amateurs.

Some important questions to ask yourself

Which instruments are more suitable for children and which for adults?

What are the main differences between refracting and reflecting telescopes?

Is my location really suitable for astronomical observing?

How much should one spend on a beginner's telescope and on more advanced equipment?

By answering these and other questions, this book outlines options and will hopefully guide you in your final choice of telescope and accessories. The chapter entitled The Basics outlines the chief optical differences between various kinds of telescopes, as well as features they have in common. This includes a brief overview of celestial characteristics that govern the design and function of telescope mountings. The next two chapters, Instruments and Accessories, describe different telescope systems and their performance characteristics in more detail. They also compare instruments and accessories available from various manufacturers, with emphasis on key design features and adaptations of each.

The book ends with a series of Appendices, including summary tables, outlining the advantages and drawbacks of all equipment described, the main questions to ask before making a final choice, and the kinds of observations possible with each type of instrument.

in mind that binoculars, refractors and reflectors are precision optical instruments. They should be constructed of quality materials and provide sharp, crisp images, with good resolution and high contrast. If all these characteristics are met, your hobby will become more than just a pleasant pastime, it will become a passionate pursuit…

Good reading !

A Brief History of Astronomical Instruments

Astronomy is truly the oldest science, as humans have cast their eyes skyward since earliest times. Although people have historically been both fearful and awed by the great celestial sphere, they have also viewed the stars as key to many mysteries. The constellations were seen as images of gods, mythological creatures and scenes from everyday life, but they also served as observational landmarks and reference points in the sky. Stonehenge, the great megalithic monument in England consisting of a vast circular array of 165 stones, was initially thought to be either a Druid temple or a Viking parliament. It was not until the clarifying work of astronomer G. Hawkins, that it became apparent that this structure was in fact a Stone Age observatory. Similar examples of ancient structures with astronomical alignments abound, including the pyramids at Giza and the temple of Ramses II in ancient Egypt, and the great stone observatories of the Mayas, to name but a few.

Lacking any sort of precise instrumentation, ancient Mediterranean mariners relied heavily on visual star sightings to navigate this great sea. The Egyptians developed a remarkably accurate agricultural calendar which was based on the morning apparition of the star Sirius and helped them keep track of the Nile's flooding cycles. Several centuries later during the Middle Ages, arab observers perfected a variety of navigational instruments, including the astrolabe, a name derived from the Greek word *astrolabos* "star-grabber". Real astronomical instruments and observatories still lay in the future, however.

Illustration of an early astronomical refractor taken from a plate published in *Selenographia*, by J. Hevelius (1647). (National Library, Paris).

The First Astronomical Telescope

For some time the invention of the astronomical refractor was erroneously attributed to Galileo Galilei. While the illustrious Italian scientist may have been the first to point a telescope at the skies (in August 1609), the invention of lenses had more modest beginnings. Although 13th century spectacle makers may have preceded them, Dutch opticians are now generally credited with inventing the telescope early in the 17th century. The first formal description of such an instrument was provided by the Italian scientist Porta in the second edition of his work *De refractione* in 1589.

Using telescopes with magnifications that never exceeded 30×,

Galileo discovered the four principal satellites of Jupiter, the surface features of the Moon and sun spots, among many other things. Unfortunately, the quality of these early telescope lenses was quite poor, particularly with respect

Antique English refractor from the 17th century (CNAM Archives, Paris).

to chromatic aberration (the splitting of light into its spectral components), a defect which could be minimized by using objective lenses of very long focal lengths. This measure was taken to extremes in the late 17th century, when astronomers built telescopes of such long focal lengths that they had to be mounted on high poles and moved by elaborate systems of ropes and pulleys. The Italian astronomer Cassini, for example, utilized a telescope system some 50 meters (about 160 ft) long!

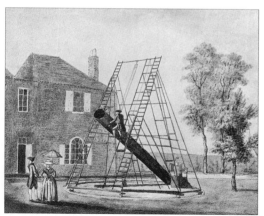

Engraving showing the great telescope of English astronomer William Herschel, installed in the garden of his house near Windsor (Library of the Paris Observatory).

The First Reflecting Telescope

One solution to the problem of chromatic aberration was the development of the reflecting telescope. Using mirrors instead of lenses, these instruments reflect light rather than refract it, and therefore avoid decomposing it into its various spectral components. In these telescopes, light is collected by a concave mirror and brought to focus. The Englishman J. Gregory first described such a system in 1661 and Isaac Newton improved on it eight years later, by introducing a small secondary mirror above the primary, which deflected the light at a right angle to an eyepiece at the side of the tube. This convenient arrangement has made the Newtonian reflector very popular among amateur astronomers today.

Telescopes Types

The most important characteristic of any telescope is the diameter or aperture of its objective, be it lens or mirror. Aperture dictates not only a telescope's ability to detect faint astronomical objects, but also determines its power of resolution, or capacity to separate two closely associated points of light. Everything else being equal (good quality optics and a solid mount), "bigger" is

Modern professional astronomers seek high altitude for the best skies, as shown here by a sunrise at the Pic de Châteaurenard observatory in the Queyras mountains, France.

definitely "better" when it comes to telescope aperture and overall performance.

Telescope performance is also affected by several external factors, especially atmospheric stability, transparency and temperature. For example, star "twinkling", an effect caused by turbulence in the Earth's atmosphere and termed "seeing" by astronomers, affects image quality in any telescope. Seeing is usually good when the atmosphere is not turbulent and bad when it is. In general as well, the larger a telescope's aperture, the more negatively impacted it is by bad seeing. In addition, air currents in the telescope tube and a sharp temperature difference between the air and the optics, can also degrade the image. As a rule, relatively small aperture, closed tube refractors, capped by the objective lens at one end and an eyepiece at the other, are least affected by these elements, while large diameter reflectors with open ended tubes suffer the most:

The Schmidt telescope combines lenses and mirrors. In addition to a mirror, it also has a corrector lens at the front. Hence, like a refractor, it is a closed instrument and is less subject to tube currents. The original Schmidt design can only be used as a camera, but a derivative, the Schmidt-Cassegrain, is very popular with amateurs because it folds a long focal length into a short, lightweight tube.

During the past 60 years, astronomy has made great strides particularly through satellites, planetary probes and space telescopes. In just a few decades, the Universe has revealed many of its mysteries to both astronomers and the public at large. Consequently, astronomy enjoys universal interest and is a pastime accessible to virtually anyone. It is noteworthy that a beginner's telescope today typically has an aperture of 10 cm (4 inches) and a magnification of 150×, in sharp contrast to Galileo's modest and low-powered telescope !

The largest telescopes

The largest refractors were built over a century ago; they are the 1-meter (40-inch) at the University of Chicago's Yerkes Observatory (1897) and, in Europe, the 83-cm (33-inch) at Meudon Observatory (1896). All larger telescopes are reflectors. The 5-meter (200-inch) reflector on Palomar Mountain was the largest in the world for many years, but today it is only the 14th largest. Among those that surpass it are two 8-meter telescopes in Hawaii and five in Chile, and the 9.2-meter Hobby-Eberly telescope in Texas. Currently, the largest telescopes in the world are the twin Keck telescopes in Hawaii, each 10 meters in aperture.

THE BASICS

It is important to have at least a basic understanding of the various optical systems available before acquiring your own. For instance, what are the fundamental differences between refractors and reflectors? Are they really that important? What are the principal optical features of a Newtonian or a Schmidt-Cassegrain? How important are such factors as aperture, focal length and resolving power? Remember too, that not all telescopes are created equal. Some are designed for very specific purposes, for instance, wide-field photography or deep-sky observing. Others are better suited for high resolution lunar and planetary astronomy, solar imaging, and so forth. It is important, therefore, to be aware of both the strengths and limitations of various telescope designs, whether this applies to optical characteristics, cost, weight and portability, sensitivity to temperature and atmospheric variations, suitability for low or high magnification, and so on. By becoming as informed about these things as possible, you are likely to make a final telescope selection in tune with both your practical needs and your budget.

The April 23, 1998 conjunction of the Moon, Jupiter and Venus.

Characteristics of Binoculars

While generally considered ideal for terrestial observing, including landscapes, wildlife and sports, binoculars are no less valued by astronomers due to their compact size, simplicity and comfort (using two eyes rather than just one), brilliant images, wide field of view and overall pleasure to use.

the basics

Different Prisms

Binoculars actually consist of two small refracting telescopes mounted in parallel. A good pair will be fully adjustable to accommodate the facial and optical requirements of various users. There are two basic types of prismatic binoculars, those with "roof" prisms and those with "Porro" prisms. These differ both in design and location within the optical path, and so determine the final shape and width of the binoculars. Prisms are simply precision blocks of glass incorporated directly into the binocular's body, which keep the optical path short and generate properly oriented images. Remember that the image produced by an objective lens is always inverted and the prisms serve to correct this.

Optical elements of roof-prism binoculars

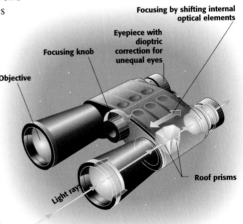

Focusing by shifting internal optical elements

Eyepiece with dioptric correction for unequal eyes

Focusing knob

Objective

Roof prisms

Light rays

Using Binoculars

Most binoculars are readily focused through a centrally located focusing knob. In Porro-prism binoculars, the knob regulates the simultaneous back and forth movement of the two eye-pieces, while in roof prism binoculars, focus is attained through movement of internal optical elements. The distinction between these two mechanisms affects not only the overall design and compactness of the two types of binoculars, but also their relative susceptibility to dust, grime and humidity. Really high end binoculars, may be sealed air tight and injected with inert gas to prevent internal condensation and associated image degradation.

Most people's eyes differ in optical acuity, even within the same individual. For this reason, binocular eyepieces can be independently adjusted through separate diopter settings. These are usually indicated by a series of engraved lines on the side of each eyepiece and can be adjusted by turning the eyepiece left or right. Once individual diopter values are determined, they can be preset to avoid having to make this correction repeatedly. This is very important not only to obtain clear focus for both eyes, but also to ensure overall viewing comfort. Unfortunately, many people are not aware that their two eyes can differ appreciably in this regard. Not only will your full enjoyment of binocular astronomy be assured through this simple adjustment, but "blurred" vision and other discomforts will be avoided. Long-eye-relief binoculars

Optical elements of Porro-prism binoculars

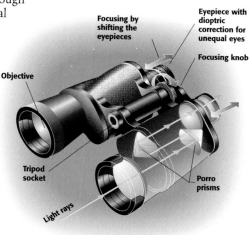

(Zeiss "B" series, Celestron Ultima, and the like) can be used comfortably with spectacles on, a necessity for those who have astigmatism.

The Next Phase in Binocular Development?

As with many other developments in optics, binoculars are also being fitted with electronic enhancements. Automatic focusing through photoelectric sensors, and electronic image stabilization features, similar to those in modern camera lenses and videocams are increasingly being adapted to binoculars. Such features stem directly from industrial developments in photography and electronic imaging.

Characteristics of Refractors

In principle, astronomical refracting telescopes are among the least complicated optical systems. Light passes through the objective lens and is bent (or refracted) into focus. This basic simplicity most likely explains why refractors were the first telescopes pointed at the stars.

Basic Optics

The heart of the refracting telescope is the objective lens at the front end of the tube assembly. The objective collects incoming light rays and forms an image inside the tube at a point called the focal plane. The eyepiece or ocular at the very back of the tube is used to both magnify and observe that primary image. In most refractors focus is achieved through a rack and pinion mechanism within the eyepiece holder.

Optical elements of the astronomical refractor

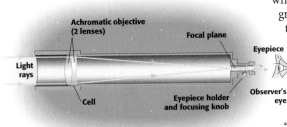

Achromatic objective (2 lenses)

Focal plane

Eyepiece

Light rays

Cell

Eyepiece holder and focusing knob

Observer's eye

Achromatic Refractors

The first refracting telescopes contained only simple biconvex objective lenses. Eyepieces likewise consisted of very simple lenses. Such lenses suffered from a variety of optical defects that were essentially insurmountable at the time. As a result, early telescopes exhibited severe chromatic aberrations (producing false coloration and image flaring) and a variety of off-axis aberrations (resulting in distorted images and a greatly reduced field of view). Fortunately, objectives in modern refractors are composed of two lens elements (typically one biconvex, the other plano-concave made of a different kind of glass), which, in combination, greatly reduce most of these optical defects. Chromatic aberration and field flatness are very much improved, earning these telescopes the title "achromatic" or color-free refractors.

Apochromatic Refractors

In an effort to produce essentially perfect optical images, refractor design and manufacture has been developed to a really fine point. By combining three and even four optical elements, as well as glass of exceptional quality and unique refractive properties, apochromatic refractors have reached that pinnacle. Among the best are refractors containing lens elements composed of high dispersion glass, rare earth elements and fluorite, which effectively eliminate all traditional optical defects and produce exquisitely sharp and contrasty images.

A small refractor.

Cautionary Notes

The optical elements in a quality telescope objective must be accurately spaced and permanently positioned in order to function to fullest capacity. For this reason, the lens elements must be housed in a finely machined metal cell which is solidly affixed to the telescope tube. To further shield the front surface of the objective lens against scratches, moisture (dew) and stray light, a solid extension tube or dew cap should be mounted outside the cell. This will extend both the longevity of the telescope's objective lens and improve its optical performance.

Things to Remember

Astronomical telescopes produce inverted images. This is no problem for astronomers, since 'up' and 'down' have little meaning in the sky. Most refractors are used with a diagonal prism, which gives an image right side up but reversed left to right. A roof prism gives a completely correct image, but high-power sharpness may suffer because of additional reflections.

Characteristics of Newtonian Reflectors

Invented in 1671 by Isaac Newton, this relatively straight forward, but elegant telescope design has today become one of the most popular and widely used. Because of its comparative simplicity and low manufacturing cost, the Newtonian reflector enjoys a significant advantage in terms of "bang for your buck" or cost per aperture, over most other types of telescopes. Its most current popular incarnation is the famed Dobsonian which has become a favorite among both novice and advanced amateurs.

A Dual Mirror System

The heart of the Newtonian reflector is its large primary mirror. This mirror must have a precisely ground, concave figure, parabolic in shape, which collects light rays and brings them to a sharp focus. It must also be mounted in a holding cell which can be adjusted mechanically. Light rays from the primary are intercepted by a much smaller secondary or diagonal mirror, which deflects them at an angle of 90° through a hole in the side of the tube. Unlike the primary, the secondary mirror must be ground and polished to an extremely flat surface, in order to reflect the light toward the observer without changing its trajectory. The secondary is held in place by a one, three or four- veined support structure called the "spider", which can also be adjusted through set screws. The

Optical elements of the Newtonian reflector

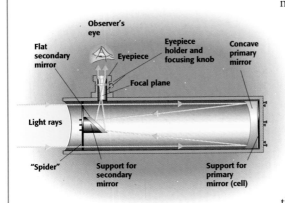

Observer's eye
Flat secondary mirror
Eyepiece
Eyepiece holder and focusing knob
Concave primary mirror
Focal plane
Light rays
"Spider"
Support for secondary mirror
Support for primary mirror (cell)

eyepiece holder is mounted at the side of the tube and may be either a rack and pinion or helical focusing device. As with any optical system, mirror surface accuracy and proper alignment are paramount in order to deliver sharp and undistorted images.

Newtonian Pros and Cons

Newtonians are simpler and less expensive than other telescopes of comparable aperture and performance. For years, until the rise of Schmidt-Cassegrains, the 150-mm (6-inch) or 200-mm (8-inch) Newtonian was the standard amateur telescope.

The Newtonian is the obvious choice if you are going to make your own mirror, since there is only one non-flat optical surface. The Newtonian is also the only design that is totally free of optical

This typical 115-mm (4.5-inch) f/8 Newtonian is a popular choice among amateur astronomers, due to its overall qualities and performance capabilities.

aberrations at the centre of the field; image quality can be (and often is) superb.

However, Newtonians are relatively bulky and hard to transport. A good equatorial mount for a Newtonian has to be massive and requires careful balancing.

Further, because the focal plane is deep within the eyepiece holder and cannot be moved, many Newtonians do not accept a camera at the prime focus, nor an image-erecting prism. Thus, Newtonians are not ideal for astrophotography or terrestrial use.

Coatings

Both the primary and secondary mirrors in reflecting telescopes are coated with an extremely thin film of aluminum making them highly reflective. The aluminum is protected by another very thin coating of quartz which confers both strength and durability to the reflective surfaces. Because they are not protected from the air, the mirrors of a Newtonian require careful cleaning every few years and recoating after a couple of decades.

Characteristics of the Schmidt, Cassegrain and Maksutov Systems

The Schmidt is an unusual telescope in amateur astronomy and really a very specialized instrument. Though less common than Newtonians, Cassegrain and Maksutov telescopes are nevertheless still widely produced and used.

The Schmidt Telescope

Invented in 1931 by B. Schmidt, this unusual instrument is not for observation at all but exclusively an astronomical camera. Its design is optimized for fast, extreme wide-field photography. It combines a short focal ratio primary mirror with a special corrector lens at the front of the tube. The latter counters the strong spherical aberrations of the primary mirror and produces ultra-sharp star images right to the edge of the field. The focal plane is located inside the telescope and 35 mm film holders are inserted through a small door at the side of the tube. Schmidt cameras are marketed to amateurs by Meade, but modern apochromatic telephoto lenses can do almost as good a job and are easier to use.

Optical elements of the Cassegrain telescope

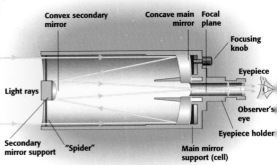

Convex secondary mirror · Concave main mirror · Focal plane · Focusing knob · Eyepiece · Observer's eye · Eyepiece holder · Main mirror support (cell) · "Spider" · Secondary mirror support · Light rays

The Cassegrain Telescope

Invented in 1672 by the Frenchman Nicolas Cassegrain, this telescope's primary mirror is mounted at the base of the tube and is pierced at the center by a large hole. Light rays collected by the primary are converged towards a secondary mirror at the front of the tube. They are then redirected backward again, to a focal point just behind the main mirror where the eyepiece holder is located. Both mirrors are mounted in mechanically adjustable cells. The secondary is held in place by a spider. Despite its obvious complexity, this compound optical system has several major advantages. By reflecting light rays backwards and forwards, a relatively long focal length telescope can be housed in a very compact tube assembly. This optical design is also quite versatile and hence suitable for a broad range of observational activities.

The Maksutov Telescope

The Maksutov, also known as the Maksutov-Cassegrain, is really a hybrid design combining elements of both refractors and reflectors. Light enters the front of the tube through a thick meniscus lens which counters the aberrations of the spherical primary mirror. As in the Cassegrain, the primary is perforated and converges the light toward a secondary mirror which returns it back through the hole in the main mirror, where it comes to focus. The Maksutov also differs from the Cassegrain with respect to its secondary mirror, which

Optical elements of the Maksutov telescope

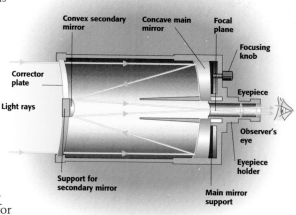

usually is an aluminized spot on the back of the meniscus lens and hence not mechanically adjustable.

A point to note

Although the optical systems of the Cassegrain and Maksutov telescopes seem to be rather different, they do have one specific characteristic in common: the image is not obtained by displacement of the eyepiece as is the case for other systems, but by displacement of the primary mirror.

The Schmidt-Cassegrain Telescope (SCT)

The Schmidt-Cassegrain telescope was developed in the 1950s by an American amateur astronomer, Thomas Johnson. This design proved so remarkable that he founded the Celestron company, which manufactures and markets these instruments to this day.

The Principle

As its name implies, the Schmidt-Cassegrain telescope (SCT) is a combination of two existing optical designs, the Cassegrain and the Schmidt. Thus we find a spherical, primary mirror at the heart of the instrument. A convex secondary mirror returns and converges the light rays through a hole in the primary mirror, and brings them to focus behind it. A special corrector plate is mounted at the front of the tube. It plays a triple role: it corrects several optical aberrations, it seals the tube tightly against dust, and it supports an adjustable secondary mirror.

Optical elements of the Schmidt-Cassegrain telescope

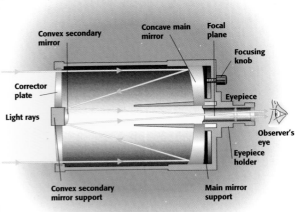

Convex secondary mirror

Corrector plate

Light rays

Concave main mirror

Focal plane

Focusing knob

Eyepiece

Observer's eye

Eyepiece holder

Convex secondary mirror support

Main mirror support

Utilization

In this type of telescope, focus is regulated by a knob at the back of the telescope. The knob is linked to an internal mechanism which moves the primary mirror back and forth. The mirror itself is factory mounted and collimated, and so cannot be adjusted independently. Collimation is adjusted at the secondary mirror and should be checked frequently. As with the Cassegrain and Maksutov, this complex telescope was designed to provide users with a powerful, large aperture instrument, in a physically compact package. These characteristics have made the SCT one of the most popular designs ever among amateur astronomers.

Celestron's Celestar-8 is a leading example of the popular 8-inch (200-mm) Schmidt-Cassegrain family of telescopes, whose optical design combines quality, performance and affordability.

Versatility at the Focal Plane

Because it focuses by moving the main mirror, the Schmidt-Cassegrain can move its focal plane through a wide range of positions; the movement of the primary is magnified by the secondary. The same is true of many Maksutov-Cassegrains. The focal plane can be placed in the eyepiece holder or as far as 30-cm (a foot) behind it.

As a result, these telescopes can take a wider range of optical accessories than any other. There is plenty of room for a diagonal, an image-erecting prism, or a camera, attached any of several ways. The focal length can be extended with a Barlow lens or reduced with a focal reducer.

What's more, the tube is short and stubby, keeping everything close to the centre of gravity, and the fork mount works without critical balancing. For these reasons, the Schmidt-Cassegrain and Maksutov-Cassegrain are popular "go anywhere, do anything" telescopes, as portable as refractors half their size.

Optical Characteristics of Various Instruments

The three major types of astronomical instruments, binoculars, refractors and reflectors, each have their own unique attributes and can be usefully compared on the basis of such common measures as aperture, focal ratio and field of view. Before making a final selection, potential buyers are advised to become familiar with these basic optical features, as well as with the strengths and limitations of various telescope designs.

Aperture

By simply doubling its diameter or aperture, a telescope will collect four times more starlight! This basic fact explains the reason for giant telescopes. In short, the larger a telescope's primary mirror or objective lens, the greater its light gathering power and ability to detect faint astronomical objects. With large aperture telescopes you can observe fainter objects more easily and clearly.

Among amateur astronomers, a telescope with a 1 meter (36–40 inch) diameter mirror pretty much falls into the "giant" category. More typically, amateur telescopes are in 200–250 mm (8–10 inch) aperture range for reflectors or Schmidt-Cassegrains, and 100 mm (4 inch) for refractors. While a 4-inch refrac-

Illustration showing the aperture and focal length of a telescope

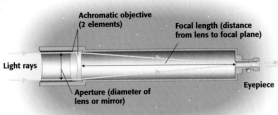

tor will provide excellent views of the moon and planets, it cannot perform as well as larger aperture telescopes with fainter deep-sky objects. It is important to note as well, that while the amount of light collected by a telescope is determined by its aperture, this has no

When selecting a telescope, aperture is very important consideration. The image at left illustrates what is visible through a small instrument. The one at right shows what can be seen with a larger aperture telescope having greater light gathering power and showing many more objects.

direct bearing on its magnifying power. On the other hand, aperture very much affects the resolving power of a telescope (see below), which in turn determines the instrument's ability to reveal diffuse planetary markings or to separate close double stars.

Focal Length

A telescope's focal length (usually abbreviated "F"), is the distance between the objective lens or mirror and the point where all light rays converge to focus. As a general rule, short focal length telescopes are preferred for wide-field, deep-sky observing (particularly with dim, diffuse objects like nebulae), while really long focal length instruments perform better for lunar planetary work. Longer focal lengths also give sharper images with less expensive types of eyepieces. The magnification of any given telescope depends on the focal length of both the objective and the focal length of the eyepiece utilized (see box below). The brightness of the visual image depends only on aperture and magnification, not focal length or f-ratio.

How to Calculate Magnification

The formula to calculate the magnification (M) of any telescope is as follows: $M = F/f$, where (F) is the telescope's focal length and (f) the focal length of the eyepiece. By international convention, all focal lengths are expressed in mm. For example, when using a 20 mm focal length eyepiece with a 900 mm telescope, the final magnification is $45\times$ ($900/20 = 45$).

Focal Ratio (f-ratio)

The focal ratio of an optical system like a telescope or camera lens is a numerical value obtained by dividing its focal length (F) by the aperture or diameter (d) of its objective lens or primary mirror. For example, a telescope of 1000 mm focal length with an objective lens or mirror 100 mm in diameter, has a focal ratio of 10. This value maybe shown as $F/d = 10$, or simply f/10. The following symbols are also used commercially to indicate the basic telescope features: diameter $\varnothing$, focal length F, and focal ratio (F/d). The table below summarizes some recommended f-ratios for various types of observing.

Resolving Power

The resolving power of a telescope is its capacity to distinguish two closely

Choosing Telescopes by Focal Ratio

Focal Ratio	Types of Observation
less than 6	deep-sky
6–10	deep-sky & lunar/planetary
greater than 10	lunar/planetary

associated points. For example, the resolving power of the human eye is approximately 1' (arc minute) (or a length of 1 cm at 34 meters). The resolving power of any optical system is directly related to its aperture; the larger the diameter of a telescope's objective lens or mirror,

Resolving Power and Atmospheric Turbulence

The following formula can be used to estimate a telescope's effective resolving power (R) under different seeing conditions; (D) is the objective or mirror diameter in millimeters:

$R = 12/D$ (under excellent seeing with little or no atmospheric turbulence)

$R = 30/D$ (under poor seeing or considerable atmospheric turbulence)

the better its resolution. The caveat to this, however, is that other factors also come into play. For example, atmospheric turbulence can vary from location to location, from one observing session to the next, or even during the same session. Excessive turbulence (or bad seeing) can greatly limit the actual resolving power of any telescope, but smaller instruments may actually perform better than large aperture telescopes under those conditions. It is important, therefore, to factor in seeing when telescopic resolution is being considered (see insert box above).

Magnitude

Ancient astronomers placed all stars into six naked-eye categories; the brightest were termed 1st magnitude, the next brightest 2nd , and so on, all the way to 6th magnitude, for barely

Diagrammatic representation comparing the resolving power of the eye with that of telescopes up to 8 inches (200 mm) in diameter

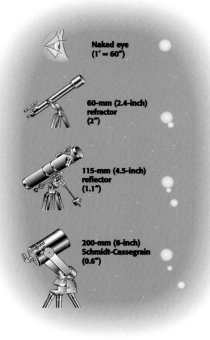

Naked eye
(1′ = 60″)

60-mm (2.4-inch) refractor
(2″)

115-mm (4.5-inch) reflector
(1.1″)

200-mm (8-inch) Schmidt-Cassegrain
(0.6″)

A yellow star, for example, will appear brighter to the eye than it does on old-fashioned blue-sensitive film. This shows that stellar magnitude determinations are to some degree dependent on the type of detector employed. For these reasons, magnitudes seen by the eye through a telescope are termed *visual magnitude*, those recorded on film are *photographic magnitude*s, while the actual or real brightness of stars is called *absolute magnitude*.

Magnification

The magnification of any telescope depends on the eyepiece or ocular being utilized. Magnification is calculated by dividing the focal length of the eyepiece (usually engraved on it) into the focal length of the telescope. It's obviously a good idea to have a selection of eyepieces of different focal length available, and to cover a range of magnifications from low, to medium, to high.

Before buying and using any eyepieces, it's best to become familiar with the various kinds available as well as with magnifications most suitable for the type of observing planned. Some magnification levels may be way too high, too low, or simply not appropriate for the telescope you are using (see insert box overleaf).

visible stars. The modern notion of stellar magnitude was introduced by Norman Pogson, who assigned each step in magnitude at 2.512 times brighter or dimmer. Using CCD (Charge Couple Device) technology, the largest modern telescopes can detect stars as faint as 30th magnitude! Our eyes do not have the same color sensitivity as photographic film.

The Field of View

One of the most important considerations in selecting an eyepiece is its field of view. In general, the size of the field of view is inversely proportional to magnification achieved. In practice, this means that it is useless, for example, to observe faint nebulae with a high power eyepiece, when a wide field and low magnification will provide brighter, more pleasing views. That is also why most telescopes are equipped with a wide-field, low power finder or spotting scope, which helps pinpoint the instrument to any spot in the sky.

When referring to eyepieces, the term "apparent field of view" is used. The size of the apparent field of view (in degrees) is determined by several characteristics, including the design or type of ocular, as well as its size and other physical parameters. The latter information is usually engraved on the eyepiece, next to the focal length specifications. The "actual field of view" is the term given to the real field of view delivered by a given eyepiece at the telescope. This value can be obtained by dividing the apparent field of view of the ocular by the final magnification attained. For example, let's use two eyepieces of very similar focal length (25 mm and 24.5 mm), but with widely different apparent fields of view (45° and 72°, respectively), and a telescope of 2000 mm focal length.

The two eyepieces have about the same magnification, $2000/25 = 80\times$ and $2000/24.5 = 81\times$ respectively. But the true fields are very different, $40°/80 = 0.56°$ for the narrow-field eyepiece and $72°/81 = 0.88°$ for the wide one. The moon, 0.5° in

How to calculate the maximally and minimally useful magnification of a telescope

The minimally useful magnification of any telescope can be calculated by dividing its aperture (in mm) by the number 7. Magnifications lower than this waste light because the beam of light emerging from the eyepiece (the *exit pupil*) is wider than the pupil of the eye. The best power for deep-sky observing is usually around a quarter to half of the aperture in mm. The power that reveals the most detail on planets is usually about equal to the aperture in mm, while maximally useful magnification should not exceed 2.5 times that value. Beyond that, image degradation due to seeing and other factors becomes severe. By way of illustration, let us use a telescope with a 200 mm (8 inch) aperture. The minimal magnification for such an instrument is $28\times$ (200/7), its optimal magnification for planets is between 200 and $300\times$ (200×1 or 1.5), and its (absolute) maximum around $500\times$ (200×2.5). Once these values have been determined, it becomes easy to select a set of eyepieces best suited for your telescope.

Photographs demonstrating the relation between apparent field of view and magnification. The image at left shows the Moon in its entirety at 60×, while the one at right shows only a small part of it at 150×.

diameter, fills the field of the first eyepiece but has plenty of room around it in the second. Clearly, eyepieces of similar focal lengths but sharply different fields of view will provide significantly different observing experiences.

Atmospheric Turbulence

Atmospheric turbulence or seeing must be factored in at all times when using a telescope, since it can greatly affect the instrument's performance. On nights of good or excellent seeing, it is possible to observe the Moon and planets under high magnification without significant image degradation. On other occasions (which can last all night), atmospheric turbulence may be so severe that you cannot even approach optimal magnification with your telescope. So, here is a word of advice for the "morning after" you have purchased a new telescope and were not pleased … don't throw it out just yet or return it to the store. The poor images you saw on your first night of observing may have less to do with the quality of your optics, than the effects of really bad seeing. Don't let your enthusiasm wane just because you did not find that 10th magnitude galaxy last night or the planets looked fuzzy. Be patient, try again and wait for nights of really good seeing to determine just what your new scope can or cannot do.

Telescope Mounts

Good mounts are essential components in using astronomical telescopes. To fully understand and appreciate this, it is important to clearly understand some of the basic rules of how the sky moves.

Sidereal Time and Daily Sky Motion

In order to observe and keep track of the objects that move with regularity across the sky – the Sun, the Moon and the planets – ancient astronomers assumed that the stars were "fixed" in the sky against a distant "celestial sphere". We know today that the stars also move in space, or have "proper" motion. However, with few exceptions, most stars are so far from us that their motion is manifest only over very long periods of time. For practical purposes, therefore, we can still consider the stars as fixed

Not always 24 hours !

It must be remembered that one rotation of the Earth does not take exactly 24 hours, but 23 hours and 56 minutes, relative to the stars. That is known as one sidereal day and explains why a star rising at 21:00 hours (9 p.m.) one night, will rise 4 minutes earlier the next night.

against a distant backdrop or "sphere". At the same time, we are aware that the constellations move slowly westward as the night progresses, an indication that the Earth itself is rotating. This is the daily motion of the sky, completed over a cycle of about 24 hr (see insert below).

Geographic Coordinates

As we view the stars from Europe and North America, our attention is naturally drawn toward the northern half of the sky. Even after just a few hours of casual observing, it becomes obvious that some stars never rise or set in that region, but turn in circles about a fixed point in the sky. This point corresponds to the Earth's axis of rotation and is conveniently marked by a well know star, Polaris. Stars which never dip below the horizon are called "circumpolar", and form part of the circumpolar group of constellations. As we turn toward the south, we notice first that the stars rise in the east, gradually move toward the zenith (their highest

This time exposure of star trails illustrates both their apparent motion in the sky and the need for a motorized telescope mount to track them.

overhead point), and then descend toward and set in the west. When a star reaches its highest point in the sky, it is at its "meridian" for that location. In other words, it is located at the exact center of the observer's north-south line. This imaginary line is called the local meridian and corresponds to the geographic longitude of the observer.

A closer study of the northern sky shows that circumpolar stars pass through two meridians, a "superior" and an "inferior" meridian, one above and one below Polaris. If the angular elevation of a star close to the North celestial pole is measured accurately at both its superior and inferior meridians, the average of those two values will indicate the exact position of the North celestial pole above the local horizon, as well as the geographic latitude at that location. In other words, the degree of elevation of the celestial pole above the horizon is exactly equal to the geographic latitude at that point. Understanding these relationships will help you use both altazimuth and equatorial telescope mounts effectively, and also provide a fuller appreciation of the underlying celestial mechanisms.

Coordinates in the sky

The simplest way to measure positions in the sky is to use *altitude* (height above the horizon, in

Diagram illustrating stellar coordinates in declination and right-ascension

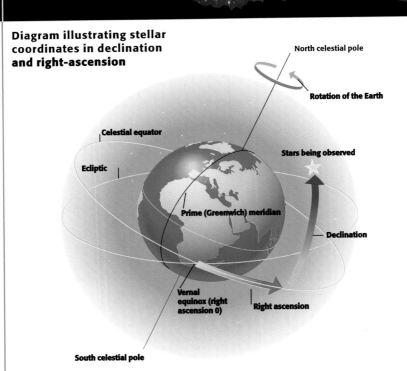

North celestial pole

Rotation of the Earth

Celestial equator

Stars being observed

Ecliptic

Prime (Greenwich) meridian

Declination

Vernal equinox (right ascension 0)

Right ascension

South celestial pole

degrees) and *azimuth* (direction, 0° = north, 90° = east, 180° = south, 270° = west). The altitude and azimuth of a celestial object change constantly because of the Earth's rotation.

For that reason, the positions of stars are normally measured in *right ascension* and *declination*. These are like longitude and latitude, respectively, on the celestial sphere, except that right ascension is measured in time units (24 hours = 360° = one full rotation). Right ascension zero is the a point known as the Vernal Equinox or First Point of Aries (♈). That is

where the Sun crosses the celestial equator every spring.

Fractional degrees are measured in minutes and seconds. Thus for example $12° 34'56'' = (12 + 34/60 + 56/3600)° = 12.582°$. Likewise, fractional right ascensions are given in hours, minutes, and seconds.

Altazimuth Mounts

The simplest telescope mounts swivel around a vertical axis, so that their movements correspond to altitude

Advantages and Disadvantages of the Altazimuth Mount

All astronomical objects appear to move in great arcs or circles across the sky. To follow this motion with an altazimuth mount, adjustments are required in both the horizontal and vertical planes, by turning both axial control knobs simultaneously. This simple mechanical process is quick and effective, but not very precise and really useful only for visual work, nothing more advanced.

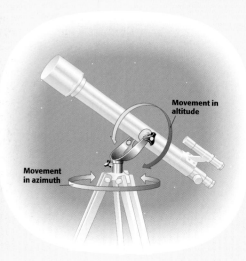

Movement in altitude

Movement in azimuth

and azimuth (see diagram above). Such telescopes are easy to aim, but their movements do not correspond to the rotation of the Earth, and until computerized mounts came along they could not hold an object in view while the Earth rotates. Nowadays, however, microprocessors make it possible for motors on both axes to track the stars and to find objects by right ascension and declination even with an altazimuth mount.

Equatorial Mounts

An equatorial mount has an axis parallel to that of the Earth, so that its movements correspond directly to right ascension and declination, and so that it can track the Earth's rotation with a single motor.

Until the advent of computerized mounts, the equatorial was the only type that could track the Earth's rotation or find objects by right ascension and declination (using graduated *setting circles*). Today, the main reason for preferring an equatorial is in order to make long-expo-sure photographs. With an altazimuth mount, although the computer can track the stars, the image slowly twists as it does so. When an equatorial mount tracks a star, the telescope and camera retain their orientation relative to the stars.

To use an equatorial mount, you must align its axis with the star Polaris, which is close to the north celestial pole. A shortcut for beginners is to point the axis straight up and use the equatorial mount as an altazimuth mount.

Different Types of Equatorial Mounts

A variety of different equatorial styles and designs exist, including the so-called German and English equatorials, as well as fork mounts and an assortment of yoke or cradle-based designs. Among amateur astronomers, the German equatorial is by far the most popular because it is particularly well-suited for smaller telescopes in the 4 to 6 inch aperture range. Although also used with larger telescopes, this type of mounting requires considerable care in terms of balancing and stabilizing these heavier instruments. For this reason, many Schmidt-Cassegrain type telescopes often utilize fork mounts, which are shorter and in many ways easier to balance and operate. No matter which type of equatorial mounting is used, however, it is

How a Computerized Mount Works

The key idea behind all computerized mounts is that two stars are sufficient to locate the whole celestial sphere. As soon as you identify two stars and align the telescope on them, it can compute the positions of all the other stars and can allow for the Earth's rotation. Until it is aligned on two stars, the telescope has to estimate the position of the sky from the observer's latitude, the date and time, and the assumption that it is level and pointed in the specified direction. Do not judge it by how well it does this; it's only making a rough estimate.

important to appreciate that for them to function as precisely and accurately as possible, they must be properly aligned, well-balanced and "broken in".

The German equatorial mount

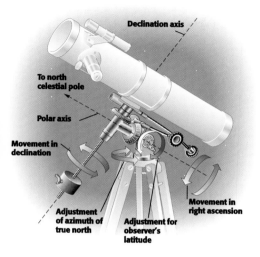

Declination axis

To north
celestial pole

Polar axis

Movement in
declination

Adjustment
of azimuth of
true north

Adjustment for
observer's
latitude

Movement in
right ascension

The equatorial fork mount

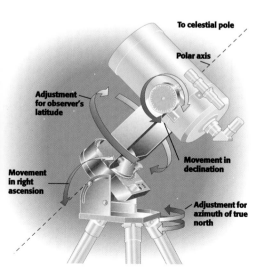

To celestial pole

Polar axis

Adjustment
for observer's
latitude

Movement in
declination

Movement
in right
ascension

Adjustment for
azimuth of true
north

THE INSTRUMENTS

These are exciting times for amateur astronomy and the selection of commercial equipment and accessories is expanding as never before. To appreciate this, one only has to look at the many telescope ads and catalogs, the choice of manufacturers and suppliers, as well as the dozens of brands of binoculars, refractors and reflectors available today. All this can be quite confusing and overwhelming for the beginner, particularly when it comes to selecting or purchasing that first piece of astronomical hardware. Should you start with binoculars, a small refractor or a full, equatorially-mounted telescope? Should you look for a very portable instrument or something larger? Which type of telescope is best for observing different astronomical objects? The following pages will assist you with these questions and, by describing the various types of commercial telescopes available, help you select instruments most suitable for your needs.

A Schmidt-Cassegrain telescope equipped with a dew cap.

Binoculars

A good set of binoculars are excellent starting equipment not only to familiarize yourself with the night sky but also to observe numerous deep-sky objects. They are particularly useful in this regard due to their brilliant images and wide fields of view.

What Can You Observe With Binoculars?

Binoculars provide wonderful views of the extended regions of the Milky Way, open star clusters, some of the brighter galaxies and of course, comets. It may seem strange, but even a modest pair of 7×50 binoculars will show the great Andromeda galaxy (M31) in all its splendor under dark skies, while the view through a much larger telescope is much more restrictive.

Basic Considerations

An optical feature called the "exit pupil" is paramount when selecting a pair of binoculars, since it greatly affects image brightness when observing at night. You will recall that the exit pupil is calculated by dividing the diameter of the objective lens by the instrument's magnification. With 7×50 binoculars, 50 mm is the diameter of the objectives and 7× is the

magnification, hence the size of exit pupil equals 50 divided by 7, or 7.1 mm. The closer this value is to 7 mm, the brighter the binocular image will appear.

Astronomical observing with a pair of binoculars can quickly become physically exhausting. A photographic tripod adapter can solve this problem and let you concentrate on observing.

One accessory to consider right away when buying binoculars is an adapter to mount them on a camera tripod. Not only will this help steady your views, but it will also lessen arm fatigue and neck cramps. These adapters are inexpensive and readily obtained, and really helpful for long-term observing. In short, when purchasing binoculars, be sure to buy a pair most suited to your personal and visual comfort.

Ordinary 7×50 and 10×50 binoculars can be used for both terrestrial and astronomical observing.

The Most Popular Binoculars: 7×50 and 10×50

About 80% of binoculars selected by amateur astronomers are of this type, since they offer the best mix of performance, weight, minimal inconvenience and price. Since optical quality for astronomical observing is less critical than for daytime viewing, entry level 7×50 or 10×50 binoculars are often sufficient. Several modestly priced brands are available from a number of manufacturers and suppliers. Their mechanical construction and overall appearance are generally also adequate. To obtain better all around performance, however, well known name brands

are preferable, including Celestron, Nikon, Olympus, Pentax and Vixen, among others. Although the cost of such binoculars will be considerably higher, their superior optical and mechanical quality will also be evident. For those desiring absolute top quality, several European brand names are available, including Leica and Zeiss. These products will provide top performance at all levels, but they are also quite costly.

If at all possible, try binoculars before you buy, to find out how well they fit your hands and eyes (and spectacles).

Giant Binoculars

Large aperture binoculars provide not only brighter images and show fainter objects than smaller aperture instruments, but they also reveal more detail in deep-sky objects like nebulae and star clusters due to markedly higher resolving power. With all binoculars in the 70–150 mm aperture range, tripod support is essential. For amateurs on a limited budget but still desiring high performance binoculars, those in the 11×70 range are probably the most appropriate, with several brands and price ranges available. Larger 12×80 binoculars are available through Vixen and others, and provide exceptional views of many astronomical objects that will captivate you for many hours on end. Under dark country skies, large binoculars will show all the Messier objects, numerous double stars, and the four brightest satellites of Jupiter.

Binoculars are well-suited for viewing extended astronomical objects, including star clusters, nebulae, large galaxies and comets (Comet Hale-Bopp is shown here).

Still larger binoculars are also available. Instruments in the 20×100 range and an exit pupil of about 5 mm, provide absolutely stunning astronomical views. These are in effect two 4-inch aperture telescopes and obviously heavy enough to require major tripod or pier support. If your budget can take it, the ultimate (25×150) giant binoculars are available from Fujinon and other manufacturers, but of course with a price and weight to match … .

Beyond a certain weight, binoculars can only be used when mounted on a photographic tripod. as shown here with a 12×80 pair.

A Binocular Revolution

Image-stabilization, a mechanism stemming directly from action photography telephoto lenses, has recently been adapted to binoculars. Canon has pioneered this, but other manufacturers are also offering this feature. These instruments use electronic sensors that act like a gyroscope through direct-drive motors to detect both horizontal and vertical motion. When movement is detected, internal processors correct the displacement by moving optical elements many times a second. The results are simply spectacular. Even after just a few moments, image shaking is inevitable with regular binoculars, due to simple muscle fatigue and eye stain. At the push of a button with image stabilizers, however, all shaking stops and your attention is focused solely on the view. These remarkable binoculars are currently available in several sizes,10×30, 12×36, etc., and are naturally quite expensive. They are well worth it, however, and add a totally new dimension to astronomical observing, be it deep-sky, the Moon or Jupiter's satellites.

Refractors for the Beginner

Many novice astronomy enthusiasts select refractors in the 50 to 60 mm aperture range as their first telescope. Such instruments are attractive because they can provide good quality views, are easy to operate and are usually quite affordable.

Solar System Observing

The Moon (see opposite) as well as many other solar system objects are natural favorites for beginners with refractors of 50 and 60 mm (2–2.4 inch) aperture. These telescopes provide pleasing views of the brighter planets, particularly the rings of Saturn, Jupiter and its moons, the phases of Venus, etc. Even though a 50-mm refractor does not show a great deal of planetary detail, the views are nonetheless pleasing to children and adults alike. A small telescope like this, however, does have limited application, particularly for adults, who will quickly outgrow its potential as their excitement and interest in this new hobby develops. A 60-mm (2.4-inch) refractor is a marked improvement over its smaller counterpart. It will reveal more planetary details as well as show the brighter deep-sky objects to advantage. These are important considerations in choosing your first telescope.

Novice astronomy enthusiasts appreciate small refractors because they are convenient and easy to operate. A Meade model is shown here, but many suppliers offer similar instruments.

Ease of Operation

Refractors of this size are extremely easy to use. Even the most inexperienced beginner will quickly learn how to operate a telescope like this if it is supplied with an altazimuth mount. Recall that these are mounts requiring only simple vertical and horizontal adjustments. Even a six year old will have little difficulty in this regard. In some instances, such small refractors may be supplied with equatorial mounts, which naturally require more refined adjustments and polar alignment. In any case, these telescopes are usually ready to use and supplied with instructions as well as several basic accessories.

Which Brands to Buy?

Numerous brand names and commercial distributors feature entry-level refractors in the 50–60 mm aperture range and with focal lengths of 600–800 mm. Virtually all these instruments are made and imported

The Moon is a favorite object for most beginners, since much detail can be seen on it even with just a 50- or 60-mm (2–2.4-inch) refractor

A Treasure Trove of Details

Any novice looking at the Moon for the first time with a 50–60-mm refractor, will be pleasantly surprised by the amount of detail visible. Even with such a small telescope, the richness of craters, mountains, channels and lava flows visible on our natural satellite is quite astounding. There is no end to the diversity of geologic features you can see, study and identify for many enjoyable evenings.

from the far East, often by the same manufacturer. The differences therefore are basically in the quality of construction, finishing details and warranties as specified by the various importers.

Numerous brand names are available that feature entry level telescopes in this size class. Some are of sufficient optical and mechanical quality, some not. In general, the more plastic components these instruments contain, the less durable their construction and longevity. Among the best know brand names in North America are Meade, Celestron, Tasco, and Orion. Most of these telescopes are made in Taiwan and China, some in Japan. The better know manufacturers will usually provide two year warranties on their products. Celestron and Orion offer a variety of small refractors of Chinese manufacture but with more robust mechanical components and good quality optics.

How Much Should I Spend?

Many small, entry level telescopes are attractively priced. A good 60-mm refractor on an altazimuth mount will retail around $100–125, and an equivalent instrument on an equatorial mount around $150–200. While less expensive telescopes are available, it is important to realize, however, that these are often little more than toys, with mostly plastic components, including lenses, and extremely flimsy mounts. Their optics will be mediocre at best and their mechanical performance no better. These "department store" instruments may be tempting but they will also disappoint you. To ensure minimum quality, therefore, it is best to avoid these and spend

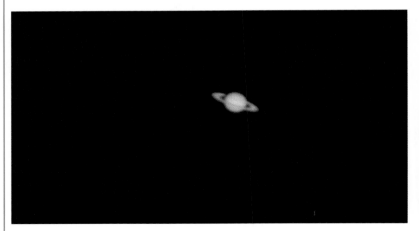

Undoubtedly, Saturn is the most beautiful planet in the Solar System. Its principal satellite, Titan, as well as its magnificent ring system are easily seen with just modest magnification.

Even with just an entry-level telescope, an equatorial mount is a great asset for astronomical observing. Not only is it easier to follow objects across the sky with it, but, if it is equipped with a motor drive to compensate for the rotation of the Earth, you can also try some basic astrophotography. The model shown here is a 60-mm (2.4-inch) refractor, equipped with a solar projection screen.

a little more on a quality product. If nothing else, get a good 20- or 25-mm eyepiece from a reputable telescope dealer; the improvement in image quality and comfort can be dramatic. You can get a "hybrid" diagonal that fits the telescope's 24.5-mm (0.965-inch) eyepiece tube and takes standard 32-mm (1.25-inch) diameter eyepieces compatible with larger telescopes. Also, make sure the mount is steady, even if you have to modify it. Many small refractors that seemed useless have found new life securely mounted on video camera tripods and the like.

Small Reflecting Telescopes

A number of small Newtonian reflectors are available in the 75–115 mm (3– 4.5 inch) aperture range, which readily outperform entry level refractors but are in the same overall price range ($100–250).

The 3-inch Newtonian

Surprisingly, a 75-mm or 3-inch reflecting telescope may not provide appreciably brighter views of astronomical objects than a small refractor. Although the 3 inch mirror is significantly larger in aperture than a 50- or 60-mm refractor, much of that apparent gain is lost due to light blockage by the secondary mirror. This "central obstruction" in a 3-inch Newtonian can reduce image brightness almost to the level of a 60-mm refractor. The 3-inch will demonstrate a small gain in resolving power over the smaller instrument and reveal more detail for instance in Jupiter's cloud band, the polar caps of Mars,

A 75-mm (3-inch) reflector.

lunar craters and Saturn's rings. With deep-sky objects, however, the only effective way to compensate for lack of aperture, is to go out to remote sites far from city lights.

Under those circumstances, a 3-inch will provide pleasing views of star clusters and the brighter nebulae like M42 in the constellation of Orion.

The 3.5-inch Newtonian

Contrary to what you might expect, a 90-mm or 3.5-inch reflecting telescope has considerably more light gathering power than its smaller (75-mm) counterpart. That "small" 15 mm increase in diameter translates into a 40% gain in area of the mirror, meaning it collects 40% more light! That, coupled with a slight increase in resolving power, makes the 3.5-inch Newtonian well suited for observing Jupiter's cloud bands and principal moons. Under good seeing, 3–4 bands will be visible on Jupiter itself, as well as the shadows of the Galilean satellites as they transit across the disk. For deep-sky observing, once again, only a good, dark location will reveal major nebulae and the brighter galaxies.

Mounts for Small Reflectors

Telescopes like these are often provided with altazimuth mounts, adjustable in both altitude and azimuth by means of micrometer

Messier 44, the "Beehive" cluster in Cancer is sufficiently bright to qualify as an ideal deep-sky object for small telescopes.

screws. Movement is controlled either through flexible cables or rigid barrel type knobs. Such mounts are generally fairly stable and work in a satisfactory manner, but they are not unfortunately of uniform quality. Fortunately, equatorial mounts are more common with this type of telescope. For the most part, these mounts are fairly stable and work well, although they are sometimes not available with motors, precluding any possibility of upgrading them in this regard. That is the case with the Celestron TC76 and similar models of far Eastern manufacture. On the other hand, some reflectors in this size class have standard Japanese equatorial mounts, similar to those supplied with the larger 115-mm or 4.5-inch Newtonians.

The 4.5-inch Newtonian

These are telescopes which are definitely at the top end of sheer beginner equipment. They are an excellent mix of aperture, price, quality and performance, and provide users ample opportunity to broaden their horizons. A 115-mm aperture mirror gathers 60% more light than a 90-mm, and will make a noticeable difference when observing nebulae, galaxies, star cluster and other deep-sky objects. Planetary details will also appear sharper and more distinct. Jupiter's famous Red Spot, the surface colors of Mars and fine lunar detail, will all be

enhanced. In short, a telescope of this size will provide many hours of enjoyable observing.

Selections Available

With so many telescopes being available in this size class, it is difficult to list all the manufacturers, brand names and models. Celestron, Tasco, Meade and Orion, are some examples, but there are many more. Once again, despite their apparent variety, all these instruments are manufactured by a handful of Japanese, Chinese and Taiwanese companies. The different

A 115-mm (4.5-inch) reflector is a very popular telescope in its class. All manufacturers offer one or models like this. The Celestron 114/900, complete with equatorial mount is a typical example.

A 115-mm aperture telescope will provide detailed views of the Moon as shown here.

brand names simply represent those of the importers who have bulk or custom ordered them. Quality control and product service will also vary accordingly.

Unfortunately, as more manufacturers enter this market, competition and lower prices have often resulted in lower quality as well. As with the very cheap refractors, the really low priced instruments in this class have many plastic components and very shaky mounts. Clearly these should be avoided. It is best to pay a little more for a quality instrument. Those offered by Meade, Orion and Celestron, for example, tend to be more rigidly constructed and equipped with fairly solid tripods and equatorial mounts. In most cases as well, those mounts can be outfitted with polar-axis alignment scopes and clock drives, both essential should you want to begin doing astrophotography.

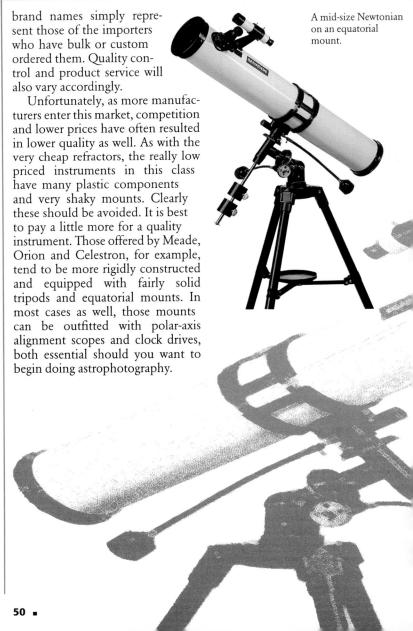

A mid-size Newtonian on an equatorial mount.

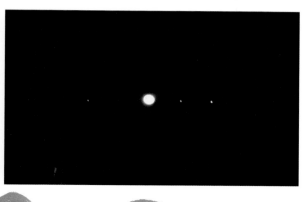

The planet Jupiter is an impressive object in any telescope, not only because of its many cloud bands, but also because its four principal satellites are easy to observe. Even a 115-mm (4.5-inch) telescope will readily show their motion in front of and behind the disk of the planet.

It Never Hurts to Ask…

A telescope is a precision optical and mechanical instrument that is best bought from a store or dealer specializing in astronomical equipment. The vendor's reputation for quality and service should always be considered before you buy. Don't hesitate therefore, to ask all and any questions you feel are necessary. Where was the instrument manufactured ? What warranties apply ? What are the dealer's return policies, and so forth. Also make sure your particular scope is properly aligned, its optics are fully collimated, and all mechanical parts move smoothly. All these precautions will avoid problems and frustrations later on.

Mid-Sized Refractors

Refractors in the 80, 90 and 100 mm (3.5 to 4 inch) range are very popular among many observers, because these are good all around telescopes. They provide crisp and detailed images of the Moon and planets, as well the brighter deep-sky objects, and they are easily used from back yards and even balconies.

Multiple Advantages

The traditional, 3–4-inch achromatic (two lens elements) refractor has many good features. It is less affected by poor seeing and sharp temperature differences than many other types of telescopes. It is almost impossible to knock out of alignment, since the objective lens is housed in a solid cell. Lastly, it has focal ratios (as short as f/9) suitable for a wide range of observing, including nebulae and star clusters.

For the Beginner

One again, a wide choice of models and brands are available in the 80-mm aperture range, but these can differ greatly in both quality and construction. For example, Celestron features an 80-mm refractor of Chinese manufacture, which is reasonably priced and a good beginner's telescope. It should be appreciated, however, that such an instrument offers little by

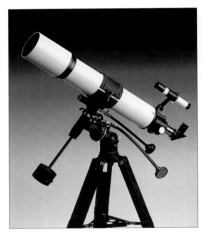

An 80-mm refractor, like this Paralux model, will easily outperform a 60-mm refractor when it comes to glimpsing planetary details.

way of upgrade, is not really equipped for high magnification and has a mount that is not rigid enough for any kind of photography.

For the More Advanced Observer

Meade Instruments, the other major American supplier, offers several interesting and reasonably priced Chinese-made refractors in the 80 to 100 mm aperture range. These are telescopes with focal lengths near 1000 mm, and are furnished with quality equatorial mounts and many accessories. Priced at between $400–600, these instruments are well suited for solar system observing. Some excellent Japanese-made telescopes are also available in this size, notably Vixen and other brand names marketed by Orion and others. These are supplied with solidly constructed German equatorial mounts in the Great Polaris series, and are available with accurate motor drives, polar alignment scopes and a variety of accessories for astrophotography. They are priced between $800–2000,

Telescope performance is determined by several criteria. This 90-mm Meade combines good optics, a stable equatorial mount and a metal tripod, thereby assuring a quality product and satisfactory observing.

depending on aperture and accessories. These are quality, well-built telescopes, and the wide range of accessories for them makes these a truly long-term investment.

Telescopes to Grow With

A properly mounted, quality 6-inch Newtonian (130–160 mm aperture range), is large and versatile enough for "serious" astronomy. In short, whether used visually or for photography, these are truly instruments you can grow with.

Great Image Quality

Compared to its smaller cousins, a 6-inch reflector represents a major gain in light-gathering power and image brightness. This is not really surprising, given that the surface area of a 6-inch mirror collects almost 50% more light than a 4.5-inch, and nearly 4 times as much as a 3-inch. This difference is clearly evident, even to the rawest novice, when observing deep-sky objects. Even the brightest usually appear as little more than faint "fuzzies" in small telescopes. In a 6-inch, however, they can be truly impressive. Due to its greater light grasp as well, a telescope of this aperture can accommodate higher magnification, meaning that globular clusters are clearly resolved into individual stars and the brighter nebulae reveal a wealth of delicate detail.

Larger mirror diameter also translates into greater resolving power. Consequently lunar and planetary observing is far more interesting. For instance, many more clearly defined bands will become apparent on

A 135-mm (5.5-inch) f/5.3 Vixen Newtonian.

Jupiter, as will surface detail on Mars (when favorably placed), including the polar caps, clouds and colorful geological features.

With a 5-inch (130-mm) or larger aperture telescope, you can begin serious observing of nebulae, star clusters and galaxies, and even deep-sky photography, realizing that most of these objects are faint and not very detailed.

Important Considerations: Your Location

Given the wide choice of options, prices and accessories available in this category of telescopes, it is important for you as a potential buyer to make some decisions before going ahead. For instance, do you live in or near a big city and do you have access to good skies? What type of observing interests you most? Is instrument portability a major factor? (see insert box overleaf).

These are important considerations, because a serious telescope of this size should really be used to fullest advantage. There is little point

in using an instrument of this aperture (or larger) under such severe light pollution so as to negate its potential. Ideally, you would at least have access to a fairly dark backyard site or park, or better still a rural location, away from glaring street lights, bill boards, or the ever-present urban sky glow and haze. An open-tubed Newtonian is also subject to air currents and sharp temperature differences, and so it is important to allow time to let the optics settle down and equilibrate with the ambient temperature. This usually takes about a half to one hour, especially during the winter or if the instrument was stored indoors.

Other Considerations

Weight and portability are also important factors to consider before purchasing a telescope of this size or larger. There is considerable variation in this regard, depending on the manufacturer, the size of the mount, accessories to be used, and so on. Typically, a relatively lightweight 6-inch Newtonian tube assembly plus mount will weigh between 40 and 50 lbs (17–25 kg). Depending on focal ratio, the tube can be anywhere from 15 to 30 inches (40–80 cm) long, factors to keep in mind for field trips, etc. At the really high-end, Takahashi offers a truly outstanding and robust 160-mm aperture Newtonian (MT160), whose tube assembly alone weighs close to 60 lbs

A 5- or 6-inch telescope with good optics will provide wonderful views of rich star fields under winter skies, including the great cluster Messier 45 in the constellation Taurus.

(30 kg), with a rock-solid mount (EM 200) to match!

The construction and finishing quality of a telescope or mount must also be considered. Many less expensive and entry-level products are attractive, reasonably well made and perform adequately. At the other extreme, products offered by Takahashi, for example, are among the very best (and most expensive) available today. Telescopes and mounts manufactured by Vixen fill the middle range, being both well

Which Scope is Most Suitable for You ?

In general, mid-range aperture telescopes with short focal ratios (f/4 to f/5) are well suited for wide-field and deep sky observing, including the brighter galaxies, star clusters and the Milky Way. Instruments of this type are not necessarily the best for lunar and planetary observing, since they are primarily intended for use with low power, wide angle eyepieces. For high magnification viewing of lunar and planetary detail, longer f-ratio (f/10 to f/13) telescopes are preferable.

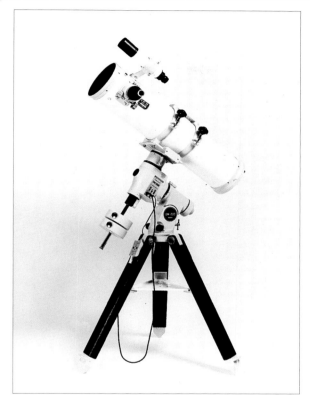

The Takahashi MT 160 telescope, shown here with the EM 200 mount, is one of the finest instruments in its class, thanks to outstanding optical and mechanical construction

products (Vixen and Takahashi) perform best.

How Much Should You Spend?

"You get what you pay for" is as true for astronomical equipment as it is for most other things. While the overall cost of mass-produced telescopes has probably never been lower nor their quality higher, the best optics and mountings do not come cheap. If it boils down to a choice between cost and aperture for example, it may be smarter to spend your limited money on a qua-

made and reasonably priced. It is well to remember that mechanical quality also influences telescope performance. Essentially all of the instruments we have described here perform well optically. However, smooth focusing mechanisms, well machined mounts and rigid tripods are equally important. Not surprisingly, here too, the more expensive

lity smaller instrument than a larger one of inferior optical or mechanical stability. As a general guide, you should invest at least $100–150 for an entry-level telescope, $200–500 for something mid-range, and $1000–3000 for a high-end instrument. The sky can easily become the limit for absolute top of the line ($5000 and up).

The 8-inch (200-mm) Telescope

Telescopes in this aperture range are very popular among amateur astronomers. They are powerful enough to provide great views both of deep-sky and solar system objects, and are highly suitable for serious astrophotography.

What can you see?

Under good seeing conditions and dark skies, a quality 8-inch telescope is a powerful instrument. Much color and detail should be visible in the bands of Jupiter and the rings of Saturn, including the elusive Encke division, and on the Moon, craters less than a mile in diameter are within reach. Among deep-sky objects, many galaxies will show distinct spiral arms, and the brighter nebulae will reveal extensive filamentous structure. Due to its popularity, a wide choice of models and optical designs are available in this class of telescopes, depending on the type of observing you intend to do. Their prices too will vary a great deal, depending on design, manufacturer, mounting and accessories. You can spend as little as $300–400 for a simple Dobsonian reflector or several thousand dollars for a top of the line compound 8-inch.

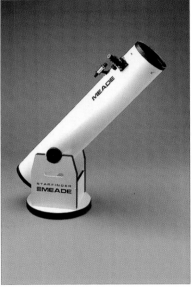

Despite their somewhat "bare bones" appearance, Dobsonians sold by Meade and Celestron, are very good telescopes for deep-sky observing, and at a very reasonable price.

Newtonian Reflectors

As we have seen, these are generally well suited for deep-sky observing, particularly those of relatively shorter focal ratios (typically around f/5). There is enough choice available in this category of instruments to fit most budgets and quality selections. The simplest type is the Dobsonian. These are widely marketed under many brand names, including Meade, Celestron and Orion, as well as by custom and speciality companies, like Coulter, Starsplitter and many others.

Dobsonian telescopes (or simply Dobs) are intended to provided maximum aperture and simplicity for minimum cost. They are usually made of simple and inexpensive materials, wood, roller cardboard tubes, plastics and Teflon, and are characterized by very simple but effective cradle-type altazimuth mounts. Their optical quality is often just average, but usually good enough for pleasing, low power and wide field views of deep-sky objects. You certainly can't beat their price per aperture factor: an 8-inch Dob can cost less than a much smaller refractor, yet deliver far brighter images. Telescopes like these are naturally limited in terms of upgrading, although Meade's Starfinder series are equatorially mounted with clock drives. Though efficient, these mounts are also rather heavy, thereby negating the main advantage of these short f-ratio telescopes.

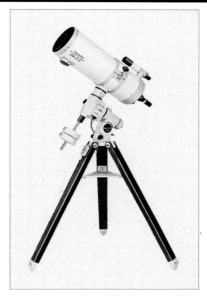

Takahashi recently added this model CN212 to its line of telescopes, a hybrid Newtonian-Cassegrain that is usable in both modes.

Vixen offers a 200-mm Newtonian of excellent quality with a fine German equatorial mount in the Great Polaris series. Its very short focal ratio (f/4) is ideally suited for wide-field, deep-sky viewing, for example the Pleiades or the extended nebulosity of NGC 7000, in Cygnus. To counteract the strong off-axis aberrations inherent in such short f-ratio optics, it is strongly advised to invest in the auxillary field-flattening lens.

Lastly, the Japanese firm Takahashi offer three different telescopes in this aperture class: a traditional Newtonian of excellent optical quality, the MT200, a strictly photo-

Due to its relatively small angular diameter, Mars requires high magnification to be seen to fullest advantage, as well as sufficient aperture to resolve much detail. A 200-mm (8-inch) telescope is well suited for this purpose.

graphic telescope, the Epsilon series, and a Newtonian-Cassegrain of unusual design. Using optical converters, the latter can be used either at f/3.9, for deep sky observing, or at f/12.4, for planetary work. All three of these are naturally at the high-end price range.

Cassegrain Telescopes

Vixen and Takahashi are two well known manufacturers specializing in this type of telescope intended mainly for lunar and planetary work. The Vixen model C200/1800 is supplied with an easy to use Great Polaris mount. The telescope also incorporates a field-flattening lens element. This makes for good, contrasty images, ideal for observing difficult planetary features like those on Mars, for example. The Mewlon Takahashi are among the best Cassegrain telescopes available commercially, with excellent optics and the EM series of German equatorial mounts. Unfortunately, they require excellent location and seeing conditions, since they are very sensitive to thermal currents, both internal and external to the optical tube assembly.

Maksutov Telescopes

Thanks to their inherently long focal lengths and f-ratios, these telescopes are best suited for lunar and planetary work. They tend to provide relatively dim images of most deep sky objects and are not suitable for wide field views or astrophotography. Although less than 200 mm in aperture, Meade features a 7-inch (178-mm) Maksutov. Optically this telescope is exceptionally good, with very contrasty images and the capacity for really high magnification. It is supplied with a fork mount, available in either equatorial or altazimuth modes and various microprocessor control options.

Schmidt-Cassegrain Telescopes (SCT)

These types of telescopes are among the most popular and widely used today. Their design is truly multi-purpose and combines sufficient aperture, focal length and power, to make it suitable for both deep-sky and solar system observing. The trade-off is that such versatility also means that the instrument does not really excel in any one category either.

The constellation Scorpius and Sagittarius contain numerous objects within grasp on a 200-mm telescope, including Messier 7 and many dark nebulae nearby.

Brilliant nebulae like Messier 42 are impressive even in small telescopes. However, a 200-mm (8-inch) telescope with more light gathering and resolving power, will permit higher magnification and provide a far more detailed and better-defined image. A short focal length, wide field eyepiece is best for this however.

At present, the two leading American manufacturers, Meade and Celestron, compete in the lucrative SCT market. Starting around $1000, these attractively priced instruments combine 8 inches (200 mm) of aperture with 2000 mm of focal length, a finder, quality eyepieces, and a stable fork mount with electronic motor drive (with Meade LX10 and Celestar 8 models). Meade actually offers a wider range of price/ features combinations than Celestron. The LX50 models come equipped with a stable fork mount and dual axis drives, but are limited in terms of photography and CCD imaging because of plastic declination gears. The LX90 and LX200 models with automatic slewing are more suitable for CCD imaging requiring accurate positioning and other advanced operations. In terms of add-ons and accessories, both Meade and Celestron offer almost unlimited possibilities.

Celestron's main strength continues to lie in the optical and mechanical quality of its products. Like the new Meade series LXD500,

Meade or Celestron?

Inevitably the question arises in the mind of every potential buyer as to which of these two brands is better. In order to address this question once and for all, it is important to realize that these two companies offer essentially identical optics in these 8-inch category telescopes. In short, for all practical purposes, it is very difficult to distinguish between them in terms of overall image quality. Your choice therefore, will ultimately be based on your budget, the type of observing and/or upgrading you are planning (for example, photography or CCD imaging), and of course the kinds of deals and services your supplier offers.

The Meade LX10 and the Celestar 8 shown here, are among the most popular 200-mm (8-inch) telescopes.

Celestron also offers its C-8 on German equatorial mounts of good quality. Manufactured in Japan, the GP and GP/D mounts are well made and perfectly adapted for astrophotography. Due to their light weight, these telescopes are easily set up and broken down, can withstand a little rough handling, and so are well-suited for the observer on the go. Celestron's more traditional, fork mounted C-8 are equally adaptable and versatile. In fact the Ultima 2000 and NextStar series with "GO TO" capabilities, combine both speed and virtually silent slewing for automatic star positioning.

Celestron's many imported products are likewise quality equipment, with good controls and solid construction, and offer warranties of up to 30 years.

Large Aperture Telescopes

"Light buckets for amateurs with dark skies" might be a good definition for this class of telescope. With telescopes of 10 inches (250 mm) aperture and up, deep-sky observing becomes a truly diverse and fascinating activity, but only at sites far removed from the light pollution of big cities and other sources.

What can you see?

As you might expect, large aperture telescopes provide the best views of virtually every type of deep-sky object, globular clusters resolved to the core, extensive detail in bright nebulae, often with a hint of color under really good conditions, and fine structure in many galaxies. Seeing permitting, large telescopes will also let you use higher magnification on many deep sky objects because of the instrument's greater light-gathering power.

The NGT 18 shown here is actually one of the best large aperture telescopes.

For lunar and planetary work, given quality optics, a dozen bands and much subtle detail will be revealed on Jupiter, Saturn's rings will be clearly resolved into their major divisions (Cassini's and Encke's), and the Moon will provide brilliant and breathtaking views (this will usually require a dark filter to avoid blinding yourself). Finally, suitably equipped, large aperture telescopes will permit you to obtain truly outstanding photographs and CCD images.

Basic Newtonians

In the large to very large aperture range, the basic Newtonian design predominates. The combination of large aperture and short focal ratios

(F/d) is particularly well suited for large, diffuse deep-sky objects, especially those not requiring high magnification. Dobsonians are hard to top in this category. They are attractively priced (typically between $500–$2000), come in a wide range of apertures (10–18 inches or 300–500 mm), and provide an opportunity to own that really large telescope of your dreams.

Several manufacturers, including Meade and Celestron, market or produce telescopes in this class. The most basic scopes of this type, however, are intended primarily for low power, visual work on deep-sky objects. They generally do this very well. Users should note, however, that the overall optical and mechanical quality of inexpensive "light buckets" like these are usually not adequate for high-powered lunar and planetary observing. This applies even to the equatorially-mounted Starfinder series from Meade.

Only a large aperture telescope will reveal detail in the nucleus of the famous Andromeda galaxy (Messier 31).

Top-of-the-Line Newtonians

For the amateur interested in a top-quality and highly versatile telescope in the big aperture category, the NGT 18 would be an excellent choice. Built in the US by JMI, this 18-inch (460-mm) reflector combines high quality optics, an open truss tube assembly and a versatile mounting complete with electronic tracking and positioning. (JMI also features smaller and larger versions of this design.) These are not inexpensive telescopes however, with the

Logistics Problems

It's pretty obvious that "big" telescopes can present "big" problems in terms of transportation and mobility. In short, there is little point in acquiring a scope of this size unless you have a vehicle to match and haul it around. That probably explains why so many amateur astronomers drive trucks, minivans and SUVs. The combination of size and weight, even when disassembled, of instruments in this class can add up to quite a challenge. If at all possible, a scope like this should be permanently in an observatory structure of some type.

NGT 18 selling at around $8000, for example. Once again, telescopes of this size are used to best advantage under dark skies, in the country, desert or mountain tops, far from city lights.

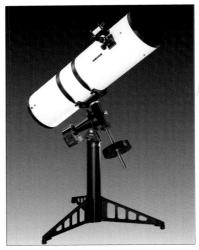

Telescopes in the Meade Starfinder series are a good compromise between aperture, basic equipment and good price. However, these instruments leave little room to upgrade for astrophotography.

Other Types of Large Telescopes

Several large aperture Schmidt-Cassegrain telescopes (SCT) are available commercially and are very popular. Ranging from 11 to 16 inches (or 239–406 mm) in diameter, these are instruments of relatively long focal length (up to 4000 mm) by virtue of their optical design. This can be frustrating when observing objects of large angular diameter, but is a real asset for high magnification, lunar and planetary work. SCTs are, however, well suited for virtually all types of astro-imaging.

Meade and Celestron are the only brands offering different sizes and models in this class of telescopes. Meade's LX200 series, complete with computerized positioning, is available in 10, 12 and 16 inch apertures (254–406 mm). If you are looking for optimal performance, the 12-inch (305-mm) Meade is an excellent choice. Its corrector plate is made of high quality BK7 glass which provides excellent image contrast. The LX200 series is fork mounted, with dual axis drives and computer controlled with a database of some 64,000 objects. This guarantees hours of enjoyable and easy observing.

Celestron also offers several models in this class of SCT, with 11 and 14-inch apertures (239–355 mm).

These are available in both fork-mounted (Ultima and NextStar series) and German equatorial versions (CM1100 and CM1400). Remember that for larger diameter telescopes the German mounts are generally preferable since they are easier to set up, use and transport. In addition, prices rise steeply with larger aperture SCTs, since these are solidly built and good quality instruments. Prices start at about $2500 and top out around $15,000 for the 16-inch Meade. Clearly, therefore, budget considerations will play a role for the typical amateur in terms of these instruments, as well as factors like expected use, accessory costs, warranties and support services, etc.

Eta Carinae, a beautiful nebula in the Southern hemisphere, is an exceptional object in a large telescope.

Apochromatic Refractors

Like very large aperture telescopes, apochromatic refractors also fall into that "highly desired" category among amateurs, since they represent the ultimate in optical performance and image quality.

Exceptional Optics

The first thing that strikes you when using an apochromatic (APO) refractor is the exceptional clarity and contrast of the images provided. Most objects also appear brighter than expected in terms of the telescope's often modest aperture. They can also handle high magnification with ease, while remaining sharp and contrasty, and seem only minimally affected by seeing or rapid changes in temperature. These superior characteristics are due to many recent advances in lens design and manufacture, that minimize or even eliminate the aberrations and distortions inherent in traditional achromatic refractor designs. In short, residual image color is completely absent and flat fields and high contrast are the norm.

A two-element apochromatic refractor

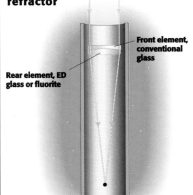

Front element, conventional glass

Rear element, ED glass or fluorite

A three-element apochromatic refractor

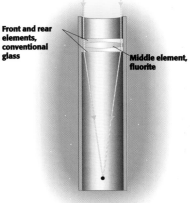

Front and rear elements, conventional glass

Middle element, fluorite

The planet Jupiter photographed with a Vixen refractor of 106 mm aperture.

Many Equivalent Choices

Five principal manufacturers currently vie in this market of excellent telescopes: Meade, Takahashi, Vixen, Tele Vue and Astro-Physics. They differ primarily in their selection of optical designs, focal ratios, types of glass and number of lens elements of their respective products. For example, some objectives incorporate only two lens elements, others three or four, some include different types of glass (ED or extra-low-dispersion, Super ED, fluorite, etc., and many proprietary formulas), focal ratios differ significantly, and so on.

Meade offers "semi" APO refractors ranging in aperture from 4 to 7 inches (100–178 mm), utilizing ED-type glass. These objectives have a mid-range focal ratio around f/9 and so fall into the multi-purpose telescope category. Images are very good overall, with high contrast and sharpness. Though generally well corrected, these lenses do retain visible chromatic aberration, particularly at higher magnifications. These are equipped with German equatorial mounts

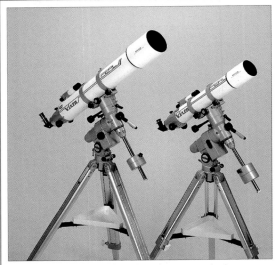

Vixen fluorite refractors have long been recognized as proven performers both optically and mechanically.

VX-102, f/9 Fluorite refractor has served as an industry standard for many years and was recently joined by an f/6.5 ED version. The fluorite element in these objectives is mounted behind the glass element in order to better protect its delicate surfaces. Optically the Vixen fluorite is similar in design and performance characteristics to the Takahashi FS series refractors. Like them, it can be used at powers up to 400× for lunar and planetary observing. Mechanically, however, both the Vixen tube assembly and mounts are somewhat less robust and well constructed.

supplied with computer drive systems similar to Meade's LX200 Schmidt-Cassegrain series.

Takahashi's FS series refractors incorporate a fluorite element in the objective lenses which produces excellent color correction even at higher magnifications. The resulting 4- and 6-inch (102- and 152-mm) telescopes provide almost perfect images, except at extremely high magnifications (e.g. at power 3–4× their diameter in mm). At f/8, these refractors also fall into the multi-purpose class of instruments. Takhashi's EM mounts are extremely rigid and of exceptionally fine quality.

Per Vixen is the first manufacturer to offer instruments of this type and quality at a moderate price. Its model

Well-Suited for Astrophotography

Unlike Meade, Takahashi and Vixen, Tele Vue features a more limited number of APO refractors. Its main product for many years was the Genesis 101. This telescope features a front element doublet made of high-performance ED glass, and another doublet situated near the image plane. The latter serves both as a field flattener and focal-reducer. As a result, this becomes a 4 inch

(100 mm) aperture precision telescope of 540 mm focal length (f/5.4) ideally suited for wide field observing and photography of extended star field and nebulosity. With long focal length eyepieces, this scope provides a field of view comparable to binoculars, providing stunning views of nebulae, star-rich Milky Way regions and comets. The machine work and mechanical construction of this instrument is also excellent. Its only drawback is that at very high magnification a hint of residual chromatic aberration is evident. Several additional Tele Vue APOs are now available in this class, including the Tele Vue 101, f/5.4 with fluorite elements and no color, as well as the TV-102, f/8.6.

The Genesis 100-mm (4-inch) f/5.4 apochromatic refractor.

Top-of-the-Line

Among all the manufacturers of APO refractors, Astro-Physics is generally regarded as the finest. Its objective lenses are true apochromatic triplets, which include Super ED glass, allowing focal ratios as short as f/5 and f/6, with effectively no residual aberration at all. Available in apertures of 105, 130 and 155 mm (4.25–6.25 inches), these telescopes are superbly machined and constructed. All elements of the tube assembly, including lens cell, rack and pinion and other accessories, are extra-large and solid. Images are excellent and withstand very high magnification under the best conditions. Unfortunately, as with Tele Vue, since production of these instruments is not assembly-line, you may have to wait several months before taking delivery.

The Price of a Small Car?

The careful construction of high-performance APO refractors can translate into "sticker-shock" when it comes to buying one. A 4-inch APO with mount can easily cost as much as a 10-inch Schmidt-Cassegrain, and the price goes up from there, with a 7-inch costing as much as small car. It all becomes a matter of priorities !

Compact Instruments

Owners of really compact telescopes are fortunate because these instruments are ideal for on the spot use. They are always available and easy to set up at a moment's notice, whether for terrestrial use (if equipped with image erector) or for astronomy.

The Price as an Index of Quality

Unfortunately, in this category of equipment "good" and "bad" are sometimes hard to tell apart, and this can make things difficult for the novice buyer when the time comes to decide. As always, it's probably best to go on the basis of your budget, since with compact optics, really good equipment is usually not available at bargain prices. As a general rule "department store" compacts or compound telescopes in the $100 to $200 range should be avoided. They invariably have mediocre optics, inferior eyepieces and unstable mounts. They may resemble good quality Schmidt-Cassegrains, for example, but they will not perform like them. Often these are sold as specials for children or as "eclipse" telescopes, but they invariably disappoint the user. It's best to look for brand names (Meade, Celestron, etc.) and be prepared to spend from $300–600 for a quality instrument of this type.

The Meade ETX is a small Maksutov of excellent quality at a moderate price.

Compact Performers

The Meade ETX and Celestron Nexstar series of compact refractors, Maksutovs and SCTs, are all quality performers in this category. The larger aperture instruments (4–5 inch, 100–125 mm) in this group will deliver good views of a variety of objects, although only the brighter deep-sky objects will be well-defined.

These instruments are also supplied either with equatorial or altazimuth mounts with electronic and/or computer controlled drives. They may also be used for photography.

Top-of-the-Line

Tele Vue manufactures some of the best-performing compact telescopes,

A compact telescope that is easy to transport can be used to advantage traveling to remote areas and photographing eclipses, for example, like this one in central America in 1998.

Ranger and Pronto. Both have similar 70 mm objective lenses and a 480 mm focal length. Using high-dispersion glass, these lenses deliver images similar to fluorite and apochromatic refractors, with bright, sharp images. The major differences between these models is in their focusing mechanisms: Ranger has a helical focuser and Pronto a classic rack and pinion and a unique carrying bag. Priced from $700 for the Ranger and $1000 for the Pronto tube assemblies, these telescopes still require either a tripod or an equatorial mount.

When combined with a camera the Tele Vue Pronto is an excellent telephoto lens for close-up terrestrial photography.

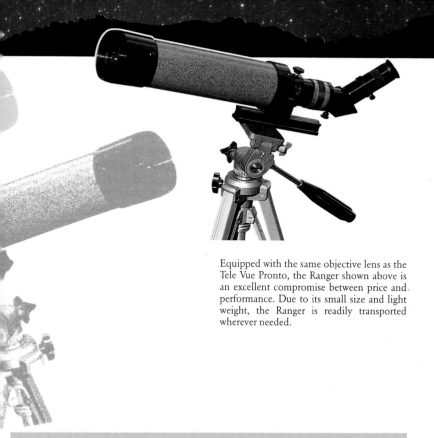

Equipped with the same objective lens as the Tele Vue Pronto, the Ranger shown above is an excellent compromise between price and performance. Due to its small size and light weight, the Ranger is readily transported wherever needed.

The Biggest of the Compacts

The Celestron 5 is among the largest aperture "compact" telescopes (the new Meade ETX 125 is also in that category) and also among the most costly ($1000–1500). The Celestron is a 5-inch (125-mm) aperture, f/10 SCT with a focal length of 1250 mm, a solid mount, with electric drive and many accessories. (The latest iteration of this telescope is the Nexstar 5, which comes complete with computer-driven go-to positioning). A 5-inch telescope of this type is big enough to provide good views of both Solar System objects and the brighter star clusters and nebulae. However, given its price, the larger 8 inch versions of both Celestron and Meade may be a better buy in the long run, since you get a lot more performance at all levels for just a little more money.

THE ACCESSORIES

Contrary to what the name might suggest, "accessories" can be absolutely essential for the full and proper function of your telescope. They can greatly affect not only the "quality" and comfort of your observing session, but also the overall range of possibilities open to you. Accessories like eyepieces, Barlow lenses, color filters, photo adapters, finder scopes, drive correctors, star charts, and so forth, are not only crucial for the proper function of your telescope, but also integral in terms of extending its capabilities beyond just the basics. For the beginner, however, the sheer number of accessories available can be quite overwhelming, and may also make it difficult to distinguish those useful items from mere gadgets.

Left: a focal reducer is shown at the focal plane of a Schmidt-Cassegrain telescope.

Eyepieces or Oculars

Few things are as central to astronomical observing as eyepieces. Acting like a magnifying glass at the focal plane of the telescope they both enlarge and transmit the image directly to the eye, thereby allowing you to visualize it.

An Overview of Eyepieces

As the last element in a telescope's optical path, the eyepiece is just as important as any of the other components. Since it forms the final image of the object you are observing it also defines all basic characteristics: sharpness, contrast, color, field of view, eye relief, etc. Every ocular should have inscribed, on its face or barrel, its name or design and focal length. These two parameters ultimately determine both the quality and the magnification of the image you ultimately perceive. That is why it is as important to select the best and most appropriate eyepieces for your particular telescope.

Modern eyepieces contain at least two or more lens elements and a barrel that fits into the eyepiece holder. The size or diameter of eyepiece barrels has been standardized into three commonly available categories. A fourth, with 27 mm diameter barrels, is no longer manufactured and was used primarily in France. So called "Japanese" eyepieces have the smallest

Calculating the Actual Field of View of an Eyepiece

The actual field of view of an eyepiece is simply a measure of the portion of sky visible through it. Expressed in degrees, this value is easily calculated by dividing the eyepiece's apparent field of view by the magnification it provides at the telescope. It is useful to compare this value to the actual angle subtended in the sky by the Moon, for example, which is approximately 0.5 degrees or 30 arc seconds. A 10-mm focal length Plossl eyepiece used with a telescope of 2000 mm focal length, gives a magnification of 200×. If the apparent field of view of this eyepiece is about 50°, its actual field of view will be 50/200 or 0.25°.

diameter (24.5 mm), "standard" American eyepieces are 1.25 inches (31.75 mm) and the largest are 2-inch (50.8 mm) diameter eyepieces.

A 24.5-mm (0.965-inch) size eyepiece is shown in an eyepiece holder of the same diameter.

Kellner (K, AR, MA) eyepieces are markedly superior to Huygens or Ramsden types. Changing to an eyepiece of this type often dramatically improves the performance of the telescope as well as the ease of using it, because you no longer have to put your eye impossibly close. In longer focal lengths, AH (achromatized Huygens) designs are also used.

The next step up comprises the Orthoscopic ("Or") and Plössl ("Pl") designs. These are four-element eyepieces that give sharp images and ample eye relief. They are available from Orion in the USA among other suppliers.

Users of Japanese-size refractors should consider getting a "hybrid" diagonal prism that fits into the 0.965-inch eyepiece tube but accepts standard 1.25-inch eyepieces.

The 0.965-inch (24.5-mm) or "Japanese" Eyepiece

This standard Japanese-size eyepiece still ranks among the most widely used in the world. It is available in an assortment of types and designs, typically with 2 or 4 lens elements. These are usually found on small, inexpensive "starter" telescopes, including 50- and 60-mm refractors and many of the 3–4-inch (75–115-mm) reflectors and are engraved with letters like "H" (Huygens), "HM" (Huygens-Mittemzwey), "R" (Ramsden) or "SR" (Special Ramsden). They are typically composed of two simple lenses, of moderate quality with a medium or small apparent field of view (~35°).While performing adequately at low powers, these oculars show their limitations as magnification increases.

The 27-mm French Eyepiece

Although no longer manufactured, this size eyepiece can be fitted to most telescopes through a custom-machined adapter. The Clavé-type eyepieces of this sort were of excellent quality and prized by many past and present observers.

Cut-away illustrations of commercial eyepieces currently available

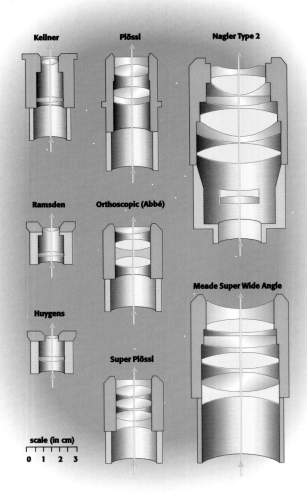

Kellner

Plössl

Nagler Type 2

Ramsden

Orthoscopic (Abbé)

Huygens

Meade Super Wide Angle

Super Plössl

scale (in cm)

0   1   2   3

The 1.25-inch (31.75-mm) Standard Eyepiece

This size eyepiece has become the world-wide standard because it offers the widest choice of design and performance characteristics. Any "serious" telescope these days, or one intended for upgrading, comes equipped with 1.25-inch eyepieces, and all major manufacturers supply them. Entry-level eyepieces in this category are typically Plössls selling in the range of $40–50. These are 4 element, well-corrected eyepieces of good to excellent quality and a 50–55° apparent field. Both Meade

1.25-inch (31.75-mm) diameter eyepieces have gradually become more popular than the more traditional 0.965-inch (24.5-mm) oculars.

and Celestron, as well as many other brands market them in focal lengths from 6.3 to 40 mm.

Top-of-the-line Plössls are available from Tele Vue and Celestron, among others, selling in the neighborhood of $150–250. In the same price range, Vixen and Takahashi offer modified Plössl eyepieces with long eye relief (20 mm regardless of focal length, a boon for spectacle wearers). The Vixen LV eyepieces use lanthanum glass and are also sold by Celestron and Orion. All of these eyepieces are multi-coated, are very sharp, and enhance the performance of any telescope.

Super-Wide Field 1.25-inch Eyepieces

Thanks to modern optical designs and availability of special glass, super-wide field eyepieces have become affordable and very popular. They give the impression that the field of view has no real boundary and make observing a very pleasant subjective experience. Selling for about $200–300, these eyepieces typically have apparent fields of 65–70° and are available from Meade (Super Wide Field series), Tele Vue (Radian and Panoptic), Pentax, and Orion and Vixen (Lanthanum Super Wide), among others. While most of these eyepieces perform very well in terms of clarity and image contrast, the Meade series does suffer from coma and other edge distortions, particularly with very short f-ratio telescopes. Pentax and Vixen eyepieces are significantly better in this regard, and the Tele Vue Panoptic series are superb. Their 70° field

Tele Vue Panoptic eyepieces are the first of a new generation of wide-field oculars. Their high optical quality guarantees exceptional images.

Meade's UWA and the Tele Vue Nagler series eyepieces are among the best available to the amateur. They combine a super-wide field of view with perfectly sharp images.

provides stunning views to the edge of the field, with high contrast and virtually no chromatic aberration. Spectacle wearers will prefer the Radians and the Pentax and Vixen eyepieces, which all have 20 mm eye relief regardless of focal length. Available in focal lengths from 15 to 35 mm these rank among the best eyepieces available today.

Ultra-Wide Field 1.25-inch Eyepieces

Ultra-wide field eyepieces (84° apparent field) provide the widest possible views attainable with any ocular. Although not all optical aberrations are as fully corrected in these eyepieces as in some other designs, their spectacularly wide views readily compensate for such minor faults. Images have high contrast and give an almost three-dimensional effect with lunar craters and bright globular cluster. Only Meade (series UWA) and the Tele Vue Nagler series market these eyepieces, in

focal lengths from 4.7 to 16 mm. Prices range from about $175–350.

The 2-inch (50.8-mm) Eyepiece

These are the largest standard-size eyepieces available commercially. A 2-inch tube is necessary to get wide fields and low powers in telescopes of high f-ratio. Eyepieces of this size are widely available in Plössl and Super Plössl designs, including the Tele Vue 55 and the Meade 56 mm, as well as wide and ultra wide formats, 27 and 35 mm Panoptics, Meade SW 40 mm, and the 20 mm Tele Vue Nagler. Some of these eyepieces, particularly 50- and 55-mm Plössls, provide enough eye relief for an elephant; you have to get used to holding your eye several inches away. In full daylight, the shadow of the secondary mirror may be visible as a dark spot in the image; this problem goes away at night when the pupil of your eye is larger.

A Range of Magnifications

As outlined previously, the magnification provided by any given eyepiece is a function of its focal length, the shorter the focal length, the higher the magnification and vice versa. Choosing an appropriate set of eyepieces for your particular telescope, therefore, depends to a large extent on what you are interested in observing.

The 8–24 mm Zoom Eyepiece

This zoom eyepiece which is imported under several brand names, comes very close in optical quality to many fixed-focal length eyepieces in the 8–24-mm range. It can be an excellent choice therefore, particularly for lunar and planetary observers who may have to change magnification quickly in response to changing atmospheric conditions and for users of computerized telescopes who must change quickly between low power, for finding and centering objects, and high power, for examining them closely. It sells for about $200.

As a rule of thumb, you should have at least three eyepieces. A low power eyepiece, for wide-field observing of deep-sky objects, a medium power eyepiece for closer views of the brighter deep-sky objects, as well as for lunar and planetary work, and a high power ocular for close-ups of the Moon and the brighter planets.

Magnification, however, should never exceed the light gathering and resolving capabilities of your particular telescope. As mentioned before, excessive magnification is of no help at all for observing faint or diffuse nebulae, nor will it do justice to lunar and planetary images which quickly deteriorate under essentially "empty" magnification. To gauge the best range of "useful" magnification for any instrument, the reader is directed to the appropriate earlier section of this book (see p. 28).

Judging an Eyepiece

Only you can decide whether a particular eyepiece works well with your eyes. Bear in mind that low-power eyepieces work well only at night, when the pupil of your eye is wide open; in the daytime, you may see dark patches darting around the field. Bear in mind also that any eyepiece has slightly more eye relief in actual use on the telescope than when you are looking at it by itself in the showroom. Even so, if you wear glasses, you will probably need eyepieces with ample eye relief.

Try an eyepiece in your own telescope or one with about the same focal ratio, and ask yourself whether the image is sharp all the way to the edges – and also whether slight blurring at the periphery is objectionable. You may prefer to see the outer regions blurred than not to see them at all.

A Variety of Accessories

In addition to eyepieces, a multitude of other accessories are available to support of your main telescope: finders, diagonals, Barlow lenses, focal reducers, etc., some of these are indispensable for optimal performance of your telescope, others simply serve to broaden its capabilities.

Finders and Pointers

Even with computerized "go to" telescopes, the finder is your most important accessory, and the better you know the sky, the more you will use it. Almost all commercial telescopes these days are supplied with finders. While these may vary in quality, they help you get started observing. Finders are simply small, low-power refractors mounted on the side of your telescope tube. Due to their wide field of view, finders help you to find and point your telescope at any objects in the sky.

Some cheaper starter telescopes, typically the 60-mm department store refractors, are equipped with 5×24 mm finders that have only simple, non-achromatic lenses and aperture stops to cut down on chromatic aberration. The devices are so poor optically as to be virtually useless, particularly with fainter objects.

To function effectively, a finder must be properly aligned with the optical axis of your telescope. This is typically done through set screws on the finder's mounting brackets or rings, allowing you to center its cross-hairs or reticle carefully on objects visible in both the finder and the main telescope. As with all astronomical refractors, the finder image will be inverted.

Hints for Using Finders

It is far easier to co-align your finder and telescope during the day, using distant objects like chimneys and flower pots, than at night, particularly if you are new at it. It may take several tries and turns of the set screws to get it just right. If your scope comes equipped with only a 5×24 finder, it is well worth replacing it with a 6×30 or better finder. Also when observing the Sun (next section) it very important to cap your finder to avoid damaging or burning the reticle, cross hairs and your face.

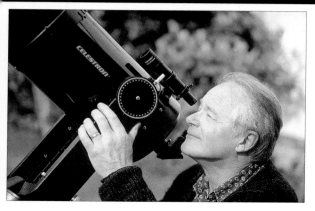

A finder is an essential accessory for any telescope. Its low power and wide field make it easy to find and pin-point an object and then center it properly in the main telescope.

Sighting pointers are relatively recent developments and very easy to use. Their operating principle is quite simple; a fine point of red light (or a series of concentric circles), are projected against the sky in the line of sight of the telescope, allowing you to pin-point it accurately in that direction. These devices have several advantages. They cover a wide field of sky at zero magnification, and are easy to align by eye due to their correctly oriented view.

Barlow Lenses

Barlow lenses can be somewhat tricky to use. They consist of a metal tube containing either single (simple Barlow), double (achromatic) or triple (apochromatic) lens elements, and are inserted into the eyepiece holder in front of the ocular. With some telescope designs there may be insufficient eyepiece travel distance to reach focus.

The sole function of a Barlow is to effectively double or triple the focal length of any telescope. In doing this, it will optically double or triple the final magnification of your instrument. For example, a 20-mm eyepiece used with a 4-inch telescope of 900 mm focal length gives a magnification of 45× (900 divided by 20). By using the same eyepiece and a two power Barlow, you can effectively double the telescope's focal length to 1800 mm and obtain a final magnification of 90×.

As with eyepieces, it is wise to get the best optical quality Barlow you can afford. Although Barlow lenses should complement your other eyepieces and might only be used occasionally, they can be a great asset, particularly for high magnification views of the Moon, planets and double stars, as well as for faint

Advantages and Disadvantages of Barlows

One of the main advantages of obtaining a Barlow lens is cost. A 2× Barlow and three or four eyepieces, for example, will provide the same range of magnification as six or eight individual oculars. Likewise for photography, a Barlow will effectively double or triple the focal length of your telescope, providing enough image scale for lunar and solar photography, as well as for eclipses.

Also, at the same magnification (such as 200×), you get to use an eyepiece with twice the focal length and twice the eye relief. But a Barlow does not change the range of powers that work well with a telescope. Thus there is no point in using a Barlow with an eyepiece that already gives high power; all you will get is a dim, blurry image.

Another disadvantage of most Barlows is that they require considerably more refocusing than is needed when you just change eyepieces. But Tele Vue now makes a Barlow lens which, when used with their eyepieces, can be inserted or removed without changing focus.

It is important to select a Barlow lens of good quality so as not to reduce the optical performance of your telescope.

variable stars and some types of photography.

Star Diagonals

Star diagonals can be very useful accessories for refractors and most Schmidt-Cassegrains are supplied from the factory with them. Diagonals, containing either a prism or a diagonal mirror, deflect light at an angle of 90 degrees and make it easier and more convenient to observe objects overhead or near the zenith. Placed in the eyepiece draw tube, diagonals do not significantly affect image brightness or quality. They give an image that is right side up but reversed left to right. Diagonals cannot be used with Newtonian reflectors. To get a completely correct image, with refractors or Schmidt-Cassegrains, use an image-erecting (roof-prism) diagonal.

Solar Projection Screen

A solar projection screen is of use mainly for group demonstrations of the Sun's disk. It consists of a small

A star diagonal is designed to minimize fatigue while observing objects high up in the sky.

screen, usually mounted on a sliding metal bar some distance behind the eyepiece, onto which the image of the Sun is projected. Although this is a relatively safe way to image the Sun with small aperture telescopes, there is always the risk that the heat generated inside the eyepiece can do damage to it. Consequently, eyepieces with multiple lens elements or cemented lens elements (as is the case in most complex modern oculars) should not be used for this purpose. Instead, it's best to use simpler 1.25-inch Huygens (H) or Ramsden (SR) type eyepieces. For similar reasons, solar projection should not be used with compound telescopes like Schmidt-Cassegrains, since the secondary mirror can be damaged. Lastly, one must be extra cautious when showing the Sun this way to young children who might be tempted to look directly into the eyepiece at the unfiltered image.

A practical way of observing sun spots is to use a solar projection screen.

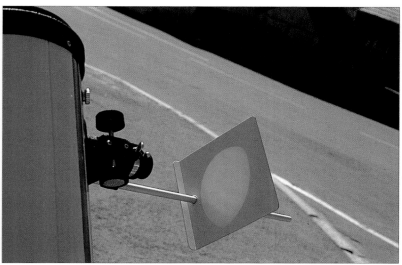

Focal Reducers

Focal reducing lenses can be very useful with certain types of telescopes, particularly Schmidt-Cassegrains. Acting in a manner opposite to that of Barlow lenses which increase the effective focal length of a telescope, this accessory shortens it, thereby reducing image magnification and increasing the field of view. For example, an f/6.3 focal reducer used with an 8-inch (200-mm) f/10 SCT will reduce its normal 2000 mm focal length to 1260 mm. As a result, a 25-mm eyepiece normally providing 80× with this instrument will now only magnify about 50×.

Focal reducers are also helpful with astrophotography, not only by providing a wider field of view but also by shortening exposure times with many objects. Again, with the above 8-inch telescope as example,

using it at f/10 and 2000 mm focal length the image scale is way too large for many objects, the full disk of the Sun or Moon, for example, during a total eclipse. A focal reducer makes it easier to frame the entire disk of the Sun or Moon on the film providing a far more pleasing picture.

Many modern focal reducers also serve as field flatteners or correctors and are particularly useful in photography because they diminish field-curvature and coma at the edge of the field. This ensures that stars will appear as points and be sharp across the entire photographic image. This is not only aesthetically more pleasing but also makes for more accurate star positioning.

Dew Shields and Lens Hoods

A dew shield or lens hood is virtually indispensable for any telescope with an exposed front lens, corrector plate or other glass element, typically found in refractors, Schmidt-Cassegrains and other compound telescopes. They are not required with Newtonian reflectors, since the long tube acts like a dew shield quite effectively.

A dew shield is really an extension of the telescope's tube which prevents or delays the formation of dew on the objective lens (particularly under humid conditions), which can quickly bring an observing session to a premature end. To be effective as long as possible, a dew shield should

A focal reducer can be used both visually and photographically. It provides a wider field of view, as well as brighter images for photographic purposes.

be at least one and a half time's as long as the objective's diameter. Finders usually come equipped with a lens hood of some type, although these are generally too short to be really effective. This is easily remedied by a construction-paper extension. A light weight cardboard shield can be used for the main instrument.

Several commercial suppliers sell dew shields designed for specific telescopes. Celestron, for example, provides light-weight dew caps which are slightly tapered at the front, thereby minimizing dew formation. Heated dew shields are also available commercially for Schmidt-Cassegrain (and other telescopes). These can prevent dew formation completely, by slightly elevating the temperature of and around the corrector plate, above the dew point, without creating any significant air turbulence. The Kendrick Dew Remover System is one of the most versatile, since the temperature is adjustable, and the controller can operate several small heaters for finders and eyepieces as well as the main lens.

Dew forms because the telescope, under the open sky, radiates heat into outer space. In so doing, it becomes cooler than the surrounding air. A dewcap retards this cooling, and a heater prevents it. Properly used, a heater may actually steady the image by eliminating the temperature difference between the lens and the air.

As a last resort, you can remove dew from a lens with an electric hair

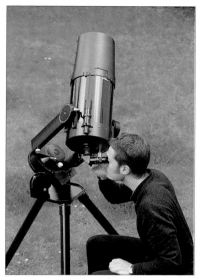

Good equipment assures good observing. This photo shows a comfortably installed observer, with a well-equipped telescope, including dew cap, star diagonal and motorized drive.

dryer (use great caution with electricity outdoors!) or a 12-volt hot-air blower designed for melting ice on car windshields. These tactics usually warm up the lens so much that the view is no longer steady. A less drastic measure is to lay a warm, dry towel across but not touching the lens for a few minutes.

Filters

A wide range of filters is available to amateur astronomers, including color filters for both observing and photography, as well as filters to minimize light pollution. In all cases, the role of filters is to enhance contrast and accentuate details of astronomical objects, by excluding or transmitting selected portions of the spectrum.

Whether they are color, neutral density or polarizing, most filters can be used directly for observing as well as for astrophotography. Filters are almost always selected for a specific purpose or object: Sun, Moon, deep-sky, etc. Over the years as well, several special anti-light pollution filters have been developed (Deep-Sky, UHR, LPR) all in an effort to minimize the increasing scourge of the amateur astronomer everywhere, the constant degradation of dark skies by artificial lights.

Solar Filters

If your telescope is a small refractor, you can view the Sun in complete safety by projecting its image onto a screen. Don't try this with a reflector, Schmidt-Cassegrain, or Maksutov, since the secondary mirror and its support can be damaged by the heat.

For this and other reasons, many observers prefer to view our star directly through the eyepiece in order to glimpse the finer details of sunspots and faculae. This requires a special filter *in front of the telescope*; filters used at the eyepiece are not safe, since they tend to crack from overheating.

Safe solar filters are made of Mylar (plastic) or glass with an aluminum coating. Suppliers include Celestron, Thousand Oaks Optical, and Baader Planetarium. If Mylar is used, it should be left somewhat loose rather than being stretched tight; two layers are required. The filter should be inspected for defects each time it is mounted on the telescope; large defects are obviously dangerous, and small pinholes admit stray light that gives a foggy image.

This turbulence is least in the early morning and, surprisingly, near sunset, when the Earth is neither warming nor cooling rapidly. Observing over a cliff or the surface of a lake is said to provide the

Colored filters are especially useful to improve visibility and contrast of planetary markings.

Filters are designed to screw directly into the eyepiece.

steadiest air; even grass is much better than hot pavement, roofs and the like.

Filters of this type are safe with telescopes of all sizes, but users of larger telescopes often mask them down to about 100 mm (4 inches) for sun viewing because the air is turbulent during the daytime.

acquire filters until they have become fairly experienced at observing the planets without them.

Whether you intend using color filters visually or photographically, it is advisable to limit your purchases to those that will be useful in terms of what you wish to observe. The following tables summarize both

Color Filters

Color glass filters can be used to great effect in both observing and photographing the planets. Their role is to enhance the relative contrast of selected planetary features such as clouds or surface details.

Although color filters are virtually indispensable to the well-trained eye of the seasoned planetary observer, they can prove disappointing to beginners who may not perceive the subtle difference in contrast or detail. For this reason, we generally do not advise novice observers to

Color	Number (#)
yellow	8
yellow-green	11
medium yellow	12
deep yellow	15
light green	56
green	58
orange	21
light red	23A
red	25A
pale blue	82A
medium blue	80A
blue	38A
violet	47

Planet	Suggested Filters	Comments
Venus	80A (medium blue)	Enhances contrast between the planet's disk and background sky and reduces the residual chromatic effects created by the extreme brightness of Venus/up to 6-inch (150-mm) apertures
	38A (blue)	Similar to above/for 6-inch (150-mm) apertures and larger
Mars	80A (light blue)	Enhances the polar caps
	23A (light red)	Reveals dark geologic surface features/ up to 8-inch (200-mm) apertures
	25A (red)	Similar to above/for 8-inch (200-mm) apertures and larger
Jupiter	80A (light blue)	Enhances the great Red Spot
	24A (light red)	Reinforces the cloud bands/up to 6-inch (150-mm) aperture
	25A (red)	Similar to above/for 6-inch (150-mm) apertures and larger
Saturn	8 (yellow)	Increases contrast between rings and the planet's disk/ up to 6-inch (150-mm) aperture
	12 or 15 (medium)	Similar to above/for 6-inch (150-mm) apertures and larger

colors and assigned filter number, as well as what specific planetary details and features they are most useful for.

Hydrogen-alpha Sun Filters

At much higher cost, a *hydrogen-alpha* (Hα) filter allows you to see prominences, flares, and other phenomena on the surface of the Sun. These filters exclude light of all wavelengths except that of hydrogen in one particular ionization state.

The filter system consists of a red pre-filter, mounted in front of the

Photo showing an H-α solar filter and its accompanying red pre-filter.

objective, and a multi-layer interference filter mounted near the focal plane. The views are truly spectacular, but the equipment costs

$800 or more. One supplier is Coronado Instruments (Ballasalla, Isle of Man IM9 2AH, Great Britain; www.coronadofilters.com).

Light-Pollution Filters

Filters of this type are available mainly as 1.25-inch (31.75-mm) sizes, with 2-inch (48-mm), and typically screw into the front end of the eyepiece or the eyepiece holder. Their function is twofold: to minimize the effects of light pollution and to selectively isolate wavelengths of light emitted by select astronomical objects. The so-called Deep-Sky and UHC (Ultra High Contrast) filters are most effective against light pollution and enhance contrast of diffuse deep-sky objects. The UHC also works extremely well on diffuse nebulae under dark skies. Its effects are striking in that it diminishes star brightness while accentuating structural detail in nebulae.

As with color filters, observers also need some experience to use interference filters effectively. It is again suggested, therefore, that the beginner first gain experience observing deep-sky objects without these aids and then acquire them later. Marvelous as these filters can be, they are not a "panacea" either, and will not immediately reveal extremely faint galaxies without considerable observing experience.

For observers in or near cities, a light-pollution filter is a very practical accessory.

Which Interference Filters for which Objects?

Diffuse nebulae: UHC

Galaxies: Deep-Sky

Very faint or small nebulae: H-beta

Planetary nebulae:
    Oxygen III

Motors and Controls

Due to the Earth's daily rotation about its axis, all objects in the sky are gradually displaced in a westerly direction. This relatively slow motion, however, only becomes apparent to the naked eye after several hours of sky watching. A telescope, of course, magnifies this effect and quickly makes it evident. For this reason, a motorized drive on your telescope mount is almost essential.

Manual Slow Motion Controls

As you observe any astronomical object through a telescope without motor drive, you quickly notice objects drifting in one direction and eventually moving out of the field of view entirely. That is why most small telescope mounts come supplied with slow-motion controls or cables, which allow you to manually reposition and track the object under observation.

With typical altazimuth mounts, manual tracking can be tricky, because this has to be done in two axes simultaneously. With a properly aligned equatorial mount this becomes much easier, since manual adjustments have to be made only in right ascension.

Why Is a Motor Drive Important?

If you have a 4.5-inch (115-mm) Newtonian reflector of 900 mm focal length or larger, and even just a beginner equatorial mount, you should get a motor drive at least for right ascension. There are several good reasons for this. First of all, observing becomes far more pleasant and comfortable. Second, you can avoid frequent adjustments and the inevitable telescope vibrations. It also becomes easier to use high magnification for lunar and planetary observing, without risking displacement of the object you are viewing. Finally, you can begin experimenting with photography of the Moon, the Sun and the brighter planets.

Unfortunately, you will not be able to photograph faint objects requiring long exposures with this type of set-up. You will not be able to reach prime focus with a small reflector of this nature you really need a solid, quality mount, with accurate motors and drive correction in both right ascension and declination. As always, this becomes a simple matter of what your budget can afford.

Most commercial 80-mm refractors and 6-inch (150-mm) aperture reflectors are available with fairly stable equatorial mounts, dual-axis motor

For visual work and for photography, at least a right-ascension motor drive is recommended, using either household current or battery power.

to compensate for the rotation of the Earth and facilitate tracking celestial objects. Equatorial mounts supplied with most entry-level telescopes (e.g. 4.5-inch reflectors and the like), typically have short shafts extending from the right ascension gear assembly which are designed to accept a drive motor. Less-expensive drives of this type may only operate through household AC power, which naturally limits their mobility in the

drives, drive correctors and other options. Once again, it is best to first become thoroughly familiar and experienced with such equipment, before getting all the additional accessories. Once you do, however, many more options are open to you, including all types of photography and electronic imaging.

It's also well to remember that drive motors are specific to each type and brand of equatorial mount and are not generally interchangeable. Keep this in mind before buying any particular accesory or piece of equipment, and be prepared to sell it later on if you are not happy with it.

Motor Drives for Beginner Telescopes

These days, even beginner telescopes equipped with equatorial mounts, can be upgraded with motor drives

Declination Drive Motors

Most equatorial mounts can also be outfitted with declination axis drive motors. Remember that stars move from east to west across the sky. Corrections in declination, therefore, are really only needed to center the telescope in a north-south direction or for guiding during long-exposure photography. Such corrections are necessary if the mount is not properly polar aligned or to compensate for small, mechanical imperfections in the drive gears, effects termed "periodic error".

The RA motor assembly must be precisely made in order to minimize irregularity in the motion of the gears.

The control panel of a high-performance telescope, showing connections for various accessories and controls.

field. Today, however, they are frequently equipped with battery-powered units.

Motor Drives for Higher-End Instruments

More complex and expensive telescope mountings are virtually always equipped with motor drives and drive correctors. These may be AC-driven via a power converter, directly by batteries or through a cigarette lighter assembly off a car battery. Controlled by variable speed drive units, mounts like these are typically equipped with hand-paddle controls. These allow you to track all astronomical objects at their appropriate rates, or quickly "slew" the telescope across the sky. Thus, whether it's to pan across the lunar surface or carefully guide during long-exposure photography, you can control the movement of your telescope accordingly.

Most good drive units can also be preset to several specific tracking rates, including solar, sidereal and lunar, to compensate for slight differences in relative motion between the planets, the Moon and background stars. Due to its proximity to Earth, the Moon, for example, exhibits a proper motion of about 13° per day against the celestial sphere, and must be tracked at that rate by the telescope.

With virtually any equatorial mount, a gradual back and forth motion in a north-south direction becomes evident over time. This may be due to slight polar misalignment or to a periodic error in the mount's gears, since no drive mechanism is perfect. In most instances, this is easily corrected with the hand control paddle.

To improve tracking during long-exposure photography, many current telescope mounts are equipped with periodic error correction (PEC) electronics. These record the actual periodic error inherent in even the best gear mechanism, and then automatically compensate for that by effecting corrective signals in right ascension. PEC is not only useful for photography but also for extended observing periods by not requiring manual tracking adjustments.

Motorized Focusing

Manual focusing can be problematic at high magnification due to vibrations introduced by turning of the focusing knob. This problem can be particularly acute during astro-imaging, where precise focus is essential. Motorized focusing has helped overcome these difficulties. A small, electric motor is attached to or over the focusing mechanism of the telescope and a push-button hand paddle allows for precise and smooth back and forth movement. The speed of the motor is adjusted by a potentiometer, resulting in virtually vibration-free control.

Motorized focusing is both convenient and useful for observing and photography. However, as with drive correctors for your mount, you must buy a model designed for your particular type and brand of telescope. Typically such devises cost about $80–100.

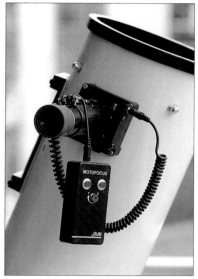

An electric focuser is useful since it eliminates any image shifting or vibrations caused by manual adjustments. This is particularly helpful when using high-power eyepieces.

Focusing is a Constant Operation

With most telescopes, frequent adjustments are required to maintain really sharp focus. There are many reasons for this. Different eyepieces and individual observers need individual adjustments. Higher magnification, like that required for lunar and planetary work, is particularly prone to changes in seeing conditions and atmospheric turbulence. As a result, fine-tuning your focus is essential.

Accessories for Astrophotography

Most amateur astronomers eventually try their hand at photographing the heavens. There is a natural progression from simply observing an astronomical object to wanting an image of it, preferably one that you have taken with you own hand and your own instruments. This can provide both a rewarding experience and a new personal perspective from you favorite pastime.

Getting an image of the Moon on film or of objects you can barely see through your telescope, like colorful nebulae and delicate galaxies, can be an exhilarating experience. To do this effectively, however, it is important to fully understand both the capabilities and limits of the instruments you have at your disposal, as well as what accessories are needed to get the job done. You will also soon discover that in the world of astrophotography, "accessories" are key and lots of them to boot!

Special photo adapters are available for virtually every type of telescope.

Principles of Astronomical Photography

In astronomical photography, your telescope simply acts as a powerful telephoto lens. For example, an 800 mm focal length telescope is equivalent to an 800 mm telephoto lens, and an 8-inch (200-mm) aperture, f/10 telescope, becomes a 2000 mm super-telephoto lens. You can very simply attach a camera to this powerful lens with two basic accessories, a T-adapter and a coupling tube. That is also the most basic arrangement to begin astrophotography. Whether you can take long or short exposures with this set-up, shoot bright or dim astronomical objects, require guiding, etc., will all hinge on which and what types of additional accessories you have at your disposal. Deciding and selecting what you really need from this panoply of options can be quite a task for beginners. Ultimately, however, practice and persistence will pay off.

The Basics

It is important before starting to take note of all items that are essential for astronomical photography. Most important among these is an equatorial mount. It must be minimally equipped with a right-ascension motor drive (and single-axis drive corrector), or an altazimuth mount with an adjustable equatorial wedge. You also need a 35 mm reflex camera, preferably an older, fully mechanical model, with a clear focusing screen and shutter with a "B" setting. It's best to avoid all modern, auto-focus lenses and electronically controlled cameras for astronomical photography, since they are totally battery-driven and therefore unsuitable for really long exposures of the night sky.

You also need a cable release to operate the shutter without shaking or vibration. Some suitable films are listed in the accompanying table. The fastest films are not the most sensitive in long exposures, a short-coming known as *reciprocity failure*. Kodak Elite Chrome 200 (E200) is a good choice. For everything except piggyback photography, you will need a so-called "T"-adapter for your specific camera brand, be it of the threaded or bayonet-type. Other

A camera adapter for piggy-back photography is shown above. This particular adapter attaches to the counter-weight shaft on a German equatorial mount.

essential adapters for you camera will depend on the type of photography planned and the objects your are shooting for.

Piggy-Back Photography

The easiest and simplest way to get started in celestial photography is to mount your camera directly onto the telescope in piggy-back fashion. This requires only a piggy-back camera adapter, which can be fitted onto the telescope tube itself or onto the counter-weight shaft of your German equatorial mount. This arrangement lets you use the camera's

Direct deep-sky photography with a SCT, requires an off-axis guider, equipped with focal reducer and an illuminated-reticle eyepiece. This combination lets you to track accurately and with minor corrections during long exposures.

own lenses and is ideal for wide-field, deep-sky photography, including constellations, star trails, meteors and the Milky Way. Since the camera is mounted in parallel with the main telescope, the latter simply serves as a guide scope to track the object you are shooting. All you do is point, lift the camera's diagonal mirror (where possible), open the shutter for a time exposure (setting B), and keep a guide star centered in an eyepiece with illuminated cross-hairs (see p.99).

Afocal Photography

The simplest way to photograph the Moon or a bright planet is simply to aim the camera, with its lens in place, into the eyepiece of the telescope. With digital and video cameras, this is the only method. The camera lens should be at maximum aperture and focused on infinity. Not all eyepieces work equally

well; experiment. The camera and telescope can even stand on separate tripods to minimize vibration.

Often confused with the afocal method, *eyepiece projection* is a technique in which the telescope has an eyepiece but the camera does not have a lens. An adapter fits around the eyepiece and screws into a 42-mm T-adapter to connect to the camera body.

Prime-Focus Adapters

To take short exposure photograph directly though a telescope, the camera is coupled to the eyepiece holder though a prime focus adapter which is screwed into the T-adapter. Prime focus adapters are available in standard 1.25-inch (24.5-mm) and 2-inch (48-mm) formats. They have a 42-mm threaded portion that mates with the camera T-adapter. With this type of arrangement, full or near-full frames of the Moon and Sun can be taken, eclipse shots of the same, and also images of larger deep-sky objects, like M42 the Orion nebula.

Off-Axis Guiders

Recording bright images on film of most deep-sky objects, typically requires exposures lasting from several minutes to an hour or more. To accomplish this most effectively you need an off-axis guider. This accessory holds the camera and is attached directly to the

Due to its brightness and large apparent diameter, the Moon is a very popular target for astrophotographers.

eyepiece holder. Most importantly, however, it also has a small prism that intercepts a portion of the light and directs it 90° toward a guiding eyepiece. The latter has illuminated cross-hairs. These allow you to focus on a guide star, and keep it centered throughout the exposure by making periodic adjustments with the drive-corrector. In this way, you can take long-exposure photographs of star clusters, nebulae and galaxies.

Some Practical Advice

Photography is a rather specialized aspect of amateur astronomy, and so you are advised not to rush into it without all the appropriate accessories, as well as with some knowledge about films, filters and the panoply of other required components. It's best to learn the basics gradually and become thoroughly familiar with all your equipment. You might practise focusing on the Moon, for instance, and guiding piggy-back exposures (which are quite forgiving of errors) before attempting more difficult long-exposures of deep-sky objects. Before proceeding, you might also seek advice from more experienced practitioners. If astronomy in general is a learning experience, you will find that success in astrophotography demands ever more practice and patience. Success in the end, however, is well worth the effort.

CCD Cameras

Astronomical imaging options have expanded enormously in recent years, thanks to the development of new technologies like CCD cameras. While these may require more complex equipment than traditional, film-based photography, the imaging possibilities are also greatly enlarged.

Operating Principles

The availability of CCD (*Charge Coupled Device*) cameras to amateurs in recent years, has greatly expanded their horizons for astronomical imaging. The basic operating principle of these devices is to capture and amplify photons emitted by an object, and to convert the information into electronic format for storage and manipulation by a computer.

Since the resultant images are stored electronically, they can be further processed and modified to improve or enhance various details and characteristics. For example, image contrast can be enhanced and/or the background can be darkened. This can be particularly helpful if you are imaging under light-polluted skies. Image processing also allows you to greatly enhance faint objects, objects that vary in magnitude or objects that move against the background stars. Among other things, this is very helpful for comet hunters and supernova searches. When coupled with appropriated imaging software (Photoshop, Photo Impact, etc.) CCD

CCD imaging is well suited for faint and delicately structured astronomical objects.

images can also be used for accurate star-positioning work (astrometry) and for generating special effects like color enhancement.

Other Advantages

In addition to their image processing capabilities, CCD cameras amplify signal electronically and are much more sensitive and efficient in capturing photons than photographic film. Consequently, exposure times can be much shorter (and therefore less affected by atmospheric turbulence) and the final image can be processed to eliminate background flaws and signal "noise".

Finally, some CCD cameras can function as automatic-guiders for a properly adapted telescope. Instead of forming an image with incoming starlight, they are linked to the telescope's drive corrector and keep it precisely on track at all times. Autoguiders replace the customary guiding eyepiece for regular astrophotography and flawlessly track objects across the sky for hours on end. Moreover, they do this more rapidly than a human operator and virtually error-free.

CCD cameras produced by SBIG are the most popular among amateur astronomers. Easy to use, these can function as both imagers and autoguiders.

Some Basic Advice

Electronic imaging has a number of drawbacks which can be an overriding factor for many amateurs. First among these is cost. CCD cameras alone can cost anywhere from about $600 to $7000, depending on complexity, performance characteristics and manufacturer. In addition, you will require a dedicated computer for the instrument, software and a variety of other operational accessories. Although both the cost and complexity of computers and CCD cameras are steadily declining, electronic imaging is still under development (and not easy to do). Moreover, most CCD cameras have extremely narrow fields of view,

and require careful pointing and alignment to capture an image in the sky. In short, this area is something many amateurs may not be ready for initially.

Only a Few Manufacturers

The number of manufacturers of CCD cameras is relatively limited. Only three American companies offer a wide range of models, Apogee, Meade and the Santa Barbara Instruments Group (SBIG). They also offer autoguiders, of which the SBIG ST-4 is the most widely used, as well as CCD camera-autoguider combinations. Cameras from Starlight Xpress in Britain are rapidly gaining a following because of their cost-effectiveness and ease of use.

Publications and Star Charts

Even just a casual visit to a large bookstore will show just how many books and publications devoted to space and astronomy are available. Magazines, reference books, encyclopedias, dictionaries, star atlases, hand books and assorted software on all aspects of amateur astronomy abound. Many are specialized and deal with various aspects of interest to amateurs, including equipment and materials.

As in previous sections of this guide, our goal here is not to critique or recommend any particular publication or product, rather it is to assist both beginners and experienced amateurs to put together a solid reference library. We have also listed in the bibliography a selection of works that may be helpful in this regard.

Planispheres

Whether just begining or more experienced, you will find a planisphere an indispensable aid to the night sky. Many different sizes and styles of these mobile star charts are available. Some indicate little more than the basic constellations, others provide considerably more information and specific detail, including the position of the planets over time and the rising and setting of various objects.

The principle underlying all planispheres is basically the same. They consist of two or more disks of cardboard and/or plastic, which can be turned relative to each other. The bottom disk typically is a complete map of the night sky, while the overlay disk is marked off in months, days and hours, and acts a window displaying the sector of sky visible at any given date and time. Planispheres are generally optimized for certain latitudes, for example, the Northern Hemisphere, Equatorial regions and the Southern Hemisphere, or areas in between. In

A wide selection of planispheres is available to suit both beginner and advanced amateurs, and all are generally easy to use.

short, with this handy devise, you can get an accurate overview of the sky visible from your location at any time.

Planisheres are particularly helpful in familiarizing yourself with the constellations, as well as for quickly revisiting any section of the sky you wish during successive observing sessions. The best way to examine a planisphere or add "any other reference material at night and not lose your dark adaptation", is to use a red-colored flash light. You can make this yourself using transparent red plastic wrap over a regular flash light, or you can purchase commercial "observing" flashlights and red plastic goggles.

The monthly star charts in magazines and in the *Cambridge Star Atlas* perform much the same function as a planisphere.

A red flash light should be used when consulting star charts during an observing session.

Star Charts and Atlases

A good star atlas complements and extends what a planisphere can do. The latter provides a quick overview of what is observable on any given night, however, any telescope will reveal objects well beyond that. For this you must have a good star chart or atlas. It will show you not only the celestial coordinates of stars, globular, nebulae and galaxies, but also their apparent size and magnitude. Information like

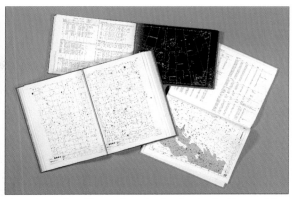

A good star atlas will complement a simpler star chart by providing more detailed information on things like double and variable stars, clusters, nebulae and galaxies, as well as magnitudes and other specific information.

that is important both in terms of what you can expect to see with your particular telescope, as well as what objects may be beyond its capabilities. Obviously it's of little use to get an atlas showing stellar magnitudes down to 14, if you are only using a 60-mm refractor.

Astronomical Ephemerides

An ephemeris is a publication giving positions of the planets and information about other constantly changing astronomical phenomena. The standard reference book of this type is the *Astronomical Almanac* published jointly by the British and U.S. governments. A wider variety of phenomena is covered more concisely in the *Handbook* of the British Astronomical Association and the *Observer's Handbook* of the Royal Astronomical Society of Canada, and in magazines such as *Sky and Telescope*.

Various ephemerides are available containing a wealth of information of astronomical events and interest (for example, the motions of Jupiter's four major moons).

As they become more experienced, and interested in specific types of observing, most amateurs will eventually select a star atlas matching those interests. Specific atlases are described on pp. 123–124. In addition, variable-star charts are available from the American Association of Variable Star Observers p. 125, and charts for many special purposes can be generated with personal computer software p. 107.

Amateurs are increasingly turning online and to astronomical software packages for access to information provided in star charts, atlases and ephemerides, and much more.

Astronomical Software and CD-ROM Reference Materials

In addition to paper publications, increasingly more computer-based applications and reference materials are becoming available to astronomers. Such programs are not only more and more comprehensive and sophisticated, but also useful as learning tools and how-to materials. Software and CD-ROM applications are replacing star charts, atlases, encyclopedias and ephemerides, and are often richly illustrated with 3-D projections and animated sequences. Star maps can be generated from such programs specific for various locations, times, horizons and other perspectives. Clearly, the possibilities are unlimited. In particular, a laptop computer running star-chart software can control a computerized telescope, and updated positions of asteroids, comets, and satellites can be downloaded from the Internet.

Many good programs of this type are available. Most of them come in more than one version, with larger star catalogues and more deep-sky and planetary information in proportion to price. Current leaders include *The Sky* (Web: http://www.bisque.com), *Starry Night* (Web: http://www.starrynight. com), and *Sky Map Pro* (Web: http://www.skymap.com). All of them use more or less the same star catalogues, mainly from Hipparcos satellite data.

Extensive astronomical reference data is also available on the World Wide Web. One good place to start is the *Sky and Telescope* web site (http://www.skypub.com). The ultimate deep-sky database for professional astronomers, also useful to advanced amateurs, is SIMBAD (http://simbad.u-strasbg.fr).

APPENDICES

The following sections are a compilation of items and references that readers will find helpful, including advice on buying used or second-hand equipment, suggestions on how to clean and maintain your telescope and accessories, how to best pack and unpack equipment during observing sessions, and listing of useful reference materials and astronomy-related addresses. You will also find an assortment of lists and tables summarizing key elements from sections of this book, and information regarding instrumentation needed for various types of observations.

Moonrise over the Saguaro Desert in the southwestern United States.

Used Equipment

An inexpensive and often very convenient way of buying telescopes and accessories is through used equipment outlets, like star party swap meets, magazine ads, online or from other amateurs. Although this can lead you to a really good deal, it is important that you make sure the used items are in good, functional condition and also what you really need or want.

Buying Used Equipment

Buying a used refractor or any other telescope, can be a very tempting pro-position, particularly for those just getting started in astronomy. Therein lies the problem: to buy used equip-ment wisely, you really should have some experience already. Only this way can you really evaluate the items you are considering and make sure they are both in good operating con-dition and precisely what you need.

That is the main reason why we advise beginners to purchase their equipment from reputable dealers or outlets that specializes in astronomy. Not only are you more likely to get good advice under those circum-stances, but also any warranties and services as required. Most telescope stores also sell used equipment. Such equipment has usually been cleaned or repaired and should be in good optical and mechanical condition. In addition, most reputable stores will also provide a three or six month warranty on any used equipment they sell.

Shining a small flashlight onto eyepiece lens surfaces will help you detect any scratches, dirt or damage to anti-reflection coatings.

Even experienced amateurs who may be quite familiar with a range of astronomical equipment, are cautioned to carefully examine any used items they are interested in. Opposite are some general guide-lines to keep in mind.

Careful examination of the reflective surface of telescope mirrors will quickly reveal any scratches or other possible damage.

What to Look for Before Buying

For starters, take a good look at the overall appearance of the instrument, the paint, the finish, scratches and dents, etc. Next, examine the optics carefully, including finders and all main telescope lenses and mirrors. It is best here to use a flash light to look for scratches, uneven coatings, small chips, and so forth. You should also examine all optical accessories this way, eyepieces, Barlows, filters, etc. Lastly, take a good look at the telescope mount. Is it fairly rigid, does it turn smoothly, are the gears clean and easy to turn in both right ascension and declination?

Once you are satisfied that the instrument passes those basic tests, try to negotiate a trial observing session, either on site or on your own. Any telescope must pass the ultimate test, image quality, under the night sky. Keep in mind, however, that atmospheric transparency or seeing can greatly influence performance.

Buying Used Accessories

When buying used eyepieces, Barlow lenses and filters, be sure to inspect them closely with the aid of a small flashlight. By shining its light on all glass surfaces, you can easily see major streaks, dust and scratches, as well as any defects in anti-reflective coatings. Similarly, camera parts and adapter tubes should inspected for internal damage or worn-out threading. Finally, ask the store if they provide any type of warranty for their used equipment and whether they recondition it in any way.

Only if You Really Need It

Used equipment is best purchased only if there is a real need and/or a significant saving over brand-new products. For example, if you are considering a larger and costlier telescope, some additional eyepieces or filters, quality photographic accessories, or if you just want to upgrade an older piece of equipment, then a look at used equipment is definitely warranted. Often you can find a real bargain or spend half as much as on a new item of comparable quality.

appendices

Some Practical Advice

Although an entire book could be devoted to practices and methods of observing, as well as equipment storage, handling and maintenance for amateur astronomers, we will only list a few common-sense rules and suggestions here.

EQUIPMENT MAINTENANCE

- The longevity of your telescope will depend on where you keep it and how well you take care of it during use.
- There was a time when the telescope mirrors had to be re-aluminized every three years. That is no longer the case thanks to protective over-coating. Newtonian optics in open tubes may have to be re-coated every twenty years or so.
- Don't be overly concerned about dust. A bit of dust on you mirror or eyepiece will not significantly affect optical performance; it's better to leave it than risk scratching delicate optical surfaces and coatings.

GETTING READY FOR AN OBSERVING SESSION

- Become thoroughly familiar with all your equipment during the day. Setting up in darkness can be difficult, frustrating and risky.
- To see stars, nebulae, and galaxies, select an observing site far away from city lights, or at least away from lights that would shine in your eyes and interfere with night vision.
* For steady views of the Moon and planets, avoid concrete patios or balconies that have been heated all day by the Sun. Set up on the lawn or other vegetation-covered surface.
- In the Northern Hemisphere, try to have a clear southern horizon, most celestial objects are south of the zenith when highest in the sky.
- Uncover your telescope and accessories a good half hour before observing, in order for it to equilibrate to ambient temperature.
- Dress warmly, even during the summer. Bring a red flash light, a good star atlas and a note book to keep a record of your observations.

DURING THE OBSERVING SESSION

- Handle all equipment gently and avoid banging into it.
- Avoid going out with too many people at once, they can cause perceptible vibrations.
- Put a dew cap on your telescope at any sign of humidity.
- Don't waste time putting back all your accessories after each use. When changing eyepieces or filters, put the one not in use in your pocket, a storage tray or a small table next to your telescope.
- Begin by observing easy objects like the Moon or planets to become familiar with your equipment, and then progress gradually to more difficult subjects. You will never see "everything" the first time out. An experienced observer may easily see six cloud bands on Jupiter on a given night, while a complete novice might barely glimpse its four major satellites! Your eyes must also become dark-adapted. Don't try to glimpse that 11th magnitude galaxy right off the bat. Chances are you won't see anything at all until you have trained your eye.
- Try to develop a logical observing program for each session. Work from west to east; the stars set in the west and you may miss the object you wanted to see in that part of the sky if you don't.

AFTER THE OBSERVING SESSION

- Don't ever wipe telescope mirrors, eyepiece or refractor lenses with a cloth, handkerchief or ordinary tissue paper. They all contain rough fibers that can easily scuff or scratch optical surfaces. Use special lens paper or a microfiber cloth for that purpose.
- Do not store eyepieces in a closed box if they are damp with dew. Leave the box partly open until morning.
- Do not cover your telescope until all dew has dissipated; bring it indoors but leave the lens caps off. On the other hand, if the telescope is cold but free of dew, install the lens caps before bringing the telescope indoors, to prevent condensation.

Advantages and Drawbacks of Various Instruments

BINOCULARS

Advantages:

- Light weight and very convenient
- Very easy to use
- Reasonably priced, at least for entry-level models
- Extremely bright images
- Very wide field of view
- Ideal for diffuse or very large objects
- Ideal for observing constellations and objects in them
- Image stabilization available in high-end models

Disadvantages:

- Low magnification (severely limits planetary observing)
- Requires a photo tripod for larger and heavier models

SMALL REFRACTORS

Advantages:

- Very affordable
- Easy to use (ideal for children)
- No difficult adjustments required
- Light weight
- Not affected much by adverse seeing
- Good lunar and planetary images

Disadvantages:

- Not much upgrading is possible
- Small aperture limits deep-sky observing
- Mounts are generally simple and lack motor drives

SMALL REFLECTORS

Advantages:

- Good value and performance for the price (especially for a 4.5-inch, 115-mm aperture)
- Sufficient aperture and resolution for both deep-sky and planetary observing

Disadvantages:

- Rather cumbersome to use
- Needs occasional adjustment after transport
- More sensitive to extraneous light (best used under dark skies)
- Takes time to thermally equilibrate (at least 2 hr)

MID-SIZED REFRACTORS

Advantages:

- Very effective under moderate conditions (both seeing and light pollution)
- Generally very good optically
- Very good for lunar and planetary work
- Sufficient size for some deep-sky work
- Usually well constructed and durable
- Can be used effectively for planetary photography
- A good all round instrument requiring little or no adjustments

Disadvantages:

- Rather costly in terms aperture/ dollar
- Very costly relative to comparable size Newtonians
- Fairly heavy and somewhat cumbersome to use
- Not enough aperture for serious deep-sky work

TELESCOPES TO GROW WITH (5–6 INCH APERTURE RANGE)

Advantages:

- Good performance/aperture/ value to build on and upgrade
- Usually good quality optics
- Aperture and f-ratio good for deep-sky observing
- Usually well and solidly built
- Highly suitable for astrophotography

Disadvantages:

- Weight and inconvenience can be appreciable
- Somewhat limited for planetary work
- Strongly affected by light pollution
- Must be temperature stabilized during winter use
- Collimation can be tricky, especially with short focal length Newtonians

THE 8-INCH (200-mm) TELESCOPE

Advantages:

- Convenient and manageable in the SCT format (Schmidt-Cassegrain)
- Good price/aperture value for SCTs
- Significant light grasp and resolution
- Almost unlimited upgrading is possible (photography, CCD imaging)
- Newtonians very good for deep-sky, SCT for lunar and planetary, at good sites

Disadvantages:

- Newtonians can be heavy and cumbersome
- Top-of-the-line instruments can be quite expensive
- Large aperture is seriously affected by light pollution
- Newtonians take time to reach thermal equilibrium, especially in winter
- Newtonians need frequent collimation adjustments, especially if transported around a lot

BIG APERTURE TELESCOPES

Advantages:

- Bigger is better ! Ideal for all kinds of observing, deep-sky and planetary if the optics are first-class
- Ideal for astro-imaging if instrument is of quality and properly equipped

Disadvantages:

- Weighty and cumbersome
- Really must have excellent skies to be used fully
- Very sensitive to light pollution & bad seeing
- Can be very expensive for quality equipment

APOCHROMATIC REFRACTORS

Advantages:

- Exceptional optical quality
- Rapid temperature equilibration
- Suitable for use under wide range of sky conditions (seeing, light polluted, etc.)

Disadvantages:

- Quite costly in terms of price/aperture
- Generally very expensive
- Larger apertures (150 mm) are quite heavy

COMPACT INSTRUMENTS

Advantages:

- Light and very convenient to use
- Optical quality often very good for more advanced (and costly) models
- Some are also good for terrestrial viewing

Disadvantages:

- Modest aperture and limited capabilities
- Limited in terms of available accessories
- Not really equal to larger telescopes, no matter how high the optical quality

Questions to help you choose

Which type or class of telescope is best for you? To help you make the right selection, answer the questions below, add their total score and decide accordingly.

How good is my observing site?	1	Mediocre
	3	Good
	5	Excellent

What is my level of knowledge?	1	Novice
	3	Intermediate
	5	Advanced

How far do I want to develop my equipment and capabilities?	1	Entry-level
	3	Mid-range
	5	Advanced

What types of observing do I want to do?	1	Everything
	3	Deep-sky
	5	Planetary

	1	less than $200
What is my maximum budget	3	$500–1000
	5	more than $1500

You are a **novice** or **intermediate** but your site is too poor or your budget too limited to invest in major equipment.

Consider a good pair of mid-size (7×50 or 10×50) **binoculars**, a mid-size (80-mm) **refractor** or 115-mm **reflector**. Get a **SCT** or a larger reflector, if your budget permits.

You are an **advanced amateur** seeking top-notch equipment with high-end performance to use at an excellent observing site.

Budget permitting get an **APO refractor** or large aperture (10-inch and up) **Newtonian** for deep-sky or **SCT** for all around work.

As an **intermediate-level** amateur, you are looking for high-quality equipment to use at a good location and are ready to spend a reasonable amount of money for it.

Consider at least a **100-mm refractor** or a **reflector over 150-mm (6-inch) aperture**.

Selections for children

Up to age 10, a pair of **7×50 binoculars or a 50–60 mm altazimuth-mounted refractor**. Above age 11, a small, **equatorially-mounted refractor or reflector** are most appropriate.

What you can observe with different size instruments

7×50 or 10×50 **Binoculars**

Moon	Principal craters, maria and major mountain chains
Sun	To be avoided at all costs!
Mars	Barely discernable disk
Jupiter	Small disk; four major moons
Saturn	Slightly elongated disk
Deep-sky	Milky Way, nebulae, star masses and some galaxies

60-mm (2.4-inch) **Aperture**

Moon	Many craters, valleys, mountains and seas
Sun	Larger sun spots
Mars	Small, featureless, orange-colored disk
Jupiter	One or two cloud bands; four major satellites
Saturn	Distinct disk and rings; one satellite (Titan)
Deep-sky	Major globular and open clusters; major nebulae; a few galaxies

115-mm (4.5-inch) **Aperture**

Moon	Many small craters; major rilles and faults; straight walls and canyons
Sun	Details in sun spots (umbra, penumbra, filaments); some faculae
Mars	Polar caps and some surface markings (not easily however)
Jupiter	Two to four cloud bands; Red Spot (with difficulty); satellite shadows across disk
Saturn	Cassini's Division when the rings are fully tilted open
Deep-sky	Fully resolved open star and partially resolved globular clusters; shape and structure in nebulae and the brighter galaxies

150-mm (6-inch) Aperture

Moon	Clearly defined small craters within larger craters; details in faults; extensive rills
Sun	Distinct structure in sun spots and surface granulation
Mars	Polar caps; all major surface features
Jupiter	Numerous cloud bands with extensive detail; Red Spot easily visible; satellite shadows across Jupiter's disk are clear and easy to see
Saturn	Major bands across disk; Cassini's Division very distinct; three to four satellites
Deep-sky	Details in open clusters; globulars almost fully resolved; much detail nebulae and galaxies

200-mm (8-inch) Aperture

Moon	Extensive detail in valleys and canyons; very small craters
Sun	Very fine structure in sun spots
Mars	Polar caps: details in all major surface features
Jupiter	Five or more cloud bands; Red Spot very easy; disks of satellites themselves during passage across Jupiter (with difficulty)
Saturn	Two or more bands across disk; Cassini's Division is easy and Encke's Division is visible during favorable inclination of rings
Deep-sky	Numerous star clusters; many faint nebulae; some structure in arms of spiral galaxies

400-mm (16-inch) Aperture

Moon	Very fine geologic detail within craters
Sun	Details within filaments
Mars	Numerous fine structures; distinct surface color changes, clouds and dust storms
Jupiter	Lots of fine detail in cloud bands, details within Red Spot; major satellite disks easily resolved
Saturn	Much detail on disk; distinct coloration; Encke's Division is easy; more than six satellites
Deep-sky	All globular clusters fully resolved into stars; detailed filaments in nebulae; details in galaxies and spiral arms easy

Bibliography

Handbooks, almanacs, general books, picture books, star atlases… Thousands of astronomy books are on the market and it is impossible to give a complete list of them. The books listed here seem to us to be suitable for the amateur astronomer interested in observing or equipment. A few older books are listed because of the continuing interest to observers.

Magazines

See page 125.

Telescopes and Equipment

Basic

Covington, Michael A. (forthcoming, 2001) *How to Use a Computerized Telescope.* Cambridge University Press. (Includes detailed instructions for LX200, Autostar, and NexStar.)

Harrington, Philip S. (1998) *Star Ware: The Amateur Astronomer's Ultimate Guide to Choosing, Buying, and Using Telescopes and Accessories.* Second edition. Wiley. (Detailed product reviews, mainly for the American market.)

Manly, Peter L. (1999) *The 20-cm Schmidt-Cassegrain Telescope: A Practical Observing Guide.* Cambridge University Press. (Now that you have the telescope, what do you do with it? This book tells you.)

Optics

Rutten, Harrie, and van Venrooij, Martin (1988) *Telescope Optics.* Willmann-Bell. (Technical.)

Suiter, Harold Richard (1994) *Star Testing Astronomical Telescopes: A Manual for Optical Evaluation and Adjustment.* Willmann-Bell.

Astrophotography

Covington, Michael A. (1999) *Astrophotography for the Amateur.* Second edition. Cambridge University Press. Web: http://www.covingtoninnovations.com/astro. (Includes film astrophotography, digital image processing, and CCDs.)

Martinez, Patrick and Klotz, Alain (1997) *A Practical Guide to CCD Astronomy.* Cambridge University Press.

Reeves, Robert (2000) *Wide-Field Astrophotography.* Willmann-Bell. (A whole book about piggybacking, taking you well beyond the obvious.)

Observing Guides

For Beginners

Consolmagno, Guy J. and Davis, Dan M. (2000) *Turn Left at Orion: A Hundred Night Sky Objects to See in a Small Telescope – and How to Find Them.* 3rd ed. Cambridge University Press.

Levy, David H. (2001) *David Levy's Guide to the Night Sky.* 2nd ed. Cambridge University Press.

Moore, Patrick (2000) *Exploring the Night Sky with Binoculars.* 4th ed. Cambridge University Press. (Also useful for getting started with a small telescope.)

Raymo, Chet (1990) *365 Starry Nights.* Simon and Schuster. (Introduces you to the whole sky, one constellation at a time.)

General Handbooks

Martinez, Patrick (ed.) (1994) *The Observer's Guide to Astronomy.* 2 vols. Cambridge University Press. (Comprehensive handbook.)

North, Gerald (1997) *Advanced Amateur Astronomy.* Cambridge University Press.

Lunar and Planetary

Dobbins, Thomas A.; Parker, Donald C. and Capen, Charles F. (1988) *Introduction to Observing and Photographing the Solar System.* Willmann-Bell.

Price, Fred W. (1994) *The Planet Observer's Handbook.* Cambridge University Press.

Rükl, Antonín (1996) *Atlas of the Moon.* Kalmbach.

Ephemerides

Astronomical Almanac. Published annually by the British and U.S. Governments. Web: http://www.usno.navy.mil; http://www.nao.rl.ac.uk. (Available in many libraries; look under government publications rather than astronomy.)

Handbook, British Astronomical Association, Burlington House, Piccadilly, London, W1V 9AG. Web: http://www.ast.cam.ac.uk/~baa. (Much shorter than *Astronomical Almanac;* for intermediate and advanced observers, mainly in Britain and Australia.)

Observer's Handbook, Royal Astronomical Society of Canada, 136 Dupont Street, Toronto, Ontario, Canada M5R 1V2. Web: http://www.rasc.ca. (For beginning, intermediate, and advanced observers anywhere in the Northern Hemisphere.)

Deep-Sky Observing

Burnham, Robert, Jr. (1978) *Burnham's Celestial Handbook.* 2nd ed. 2 vols. Dover. (Beginning to be dated, but still fascinating. An encyclopedia of thousands of interesting objects, arranged by constellations.)

Houston, Walter Scott (1999) *Deep-Sky Wonders.* Sky Publishing. (From his columns in *Sky and Telescope.*)

Luginbuhl, Christian B. and Skiff, Brian A. (1998) *Observing Handbook and Catalogue of Deep Sky Objects.* 2nd ed. Cambridge University Press. (Up-to-date information about clusters, nebulae, and galaxies.)

Malin, David, and Frew, David J. (1995) *Hartung's Astronomical Objects for Southern Telescope.* Cambridge University Press. (Update of a classic guide to objects visible from the Southern Hemisphere.)

O'Meara, Stephen James (1998) *The Messier Objects.* Cambridge University Press.

Webb, Rev. T. W. (1917) *Celestial Objects for Common Telescopes.* 2 vols. Reprinted, Dover, 1962. (A classic, still available.)

Charts and Atlases

Karkoschka, E. (1999) *The Observer's Sky Atlas.* Springer. (Very compact and condensed; handy to carry in your eyepiece box.)

Ridpath, Ian (1998) *Norton's Star Atlas and Reference Handbook.* 19th ed. Longman. (stars to magnitude 6.5 charted in a plain style; lists of interesting objects; extensive reference material.)

Sinnott, Roger W. and Perryman, Michael A. C. (1997) *Millennium Star Atlas.* 3 vols. Sky Publishing. (Giant 1548-chart atlas of stars to magnitude 11 based on Hipparcos satellite data. For use with larger telescopes.)

Tirion, Wil (2001) *The Cambridge Star Atlas.* 3rd ed. Cambridge University Press. (Ideal star atlas for the beginner or small-telescope user. A junior edition of *Sky Atlas 2000.0* plus monthly charts of the whole sky.)

Tirion, Wil; Rappaport, Barry; and Lovi, George. (2001) *Uranometria 2000.0.* 2nd ed. 2 vols. Willmann-Bell. (Stars to ninth magnitude and numerous deep-sky objects; charts are approximately A4 size, convenient for use at the telescope.)

Tirion, Wil and Sinnott, Roger W. (1998) *Sky Atlas 2000.0.* 2nd ed. Cambridge University Press. (Beautiful, colorful charts of stars to magnitude 8.5 and numerous deep-sky objects. Especially useful for planning and interpreting wide-field photographs. Also available in black and white on loose charts.)

Astronomical Science

General

Mitton, Jacqueline (2001) *Cambridge Dictionary of Astronomy.* Cambridge University Press.

Moore, Patrick (1998) *Atlas of the Universe.* Cambridge University Press. (Not just an atlas, a colorfully illustrated guide to all of astronomy, somewhat slanted toward amateur observers.)

Karttunen, H., *et al.* (eds.) (1996) *Fundamental Astronomy.* Springer. (An introductory astronomy text that goes deeper into astrophysics than most.)

Cosmology

Filkin, David and Hawking, Stephen (1998) *Stephen Hawking's Universe: The Cosmos Explained.* Basic Books.

Rees, Martin J. (2001) *Just Six Numbers: The Deep Forces That Shape the Universe.* Basic Books.

History of Astronomy

Hoskin, Michael (ed.) (1997) *The Cambridge Illustrated History of Astronomy.* Cambridge University Press.

Thurston, Hugh (1994) *Early Astronomy.* Springer. (From ancient times to Kepler.)

Useful addresses

Astronomical Equipment

Buying a telescope and any additional accessories to upgrade the instrument should not be done casually. We hope that this book has helped you in that regard, given that over 300 commercial telescopes are available today and countless more accessories. By now, you should have enough information and at least some idea of the type of instrument that interests you, and we know that sooner or later you will go out and buy it … Before you do, however, we urge you to visit a store that specializes in telescopes and astronomy, and ask as many questions you might still have.

This will provide you not only the right information, but also the advice of people who are either amateurs them-selves or at least have some experience with observing and astronomical equipment. Their advice will give you more confidence as a buyer and also provide access to accessories and other helpful information. You can also expect good service after you have made your purchases.

Magazines

Astronomy,
Kalmbach Publishing Co.,
P.O. Box 1612, Waukesha,
WI 53187, U.S.A.
Web: http://www.astronomy.com.

Astronomy Now,
P.O. Box 175,
Tonbridge,
Kent TN10 4ZY, England.
Web: http://www.astronomynow.com.

Sky and Telescope,
Sky Publishing Corp.,
P.O. Box 9111, Belmont,
MA 02178, U.S.A..
Web: http://www.skypub.com.

Organizations

American Association of Variable Star Observers (AAVSO),
25 Birch Street, Cambridge,
Massachusetts 02138, U.S.A.
Web: http://www.aavso.org.

Association of Lunar & Planetary Observers (ALPO),
c/o Mr. H. D. Jamieson,
P.O. Box 171302, Memphis,
TN 38187, U.S.A.
Web: http://www.lpl.arizona.edu/alpo.

Astronomical Society of
New South Wales,
GPO Box 1123,
Sydney 1043, NSW, Australia.
Web: http://www.asnsw.com.

Astronomical Society of Southern Africa,
PO Box 9, Observatory 7935,
South Africa.
Web: http://www.saao.ac.za.

British Astronomical Association (BAA),
Burlington House, Piccadilly,
London W1V 9AG, England.
Web: http://www. ast.cam.ac.uk/~baa.
(Has members all over the world;
publishes useful *Handbook* and
Journal.)

Royal Astronomical Society of Canada,
136 Dupont Street, Toronto, Ontario
M5R 1V2, Canada.
Web: http:// www.rasc.ca.

Royal Astronomical Society of
New Zealand,
P.O. Box 3181, Wellington, New Zealand.

Manufacturers

Apogee Instruments,
3340 N. Country Club #103,
Tucson, AZ 85716, U.S.A.
Web: http://www. apogee-ccd.com
(CCD cameras for advanced amateurs
and professionals.)

Astro-Physics,
11250 Forest Hills Road, Rockford,
IL 61115, U.S.A.
Web: http://www. astro-physics.com.
(Premium-quality refractors designed
for photography.)

Celestron International,
2835 Columbia Street, Torrance,
CA 90503, U.S.A.
Web: http://www. celestron.com.

(The original Schmidt-Cassegrain
telescope manufacturer. Now makes
telescopes and accessories of all
kinds.)

JMI,
810 Quail St., Unit E, Lakewood,
CO 80215.
(Large telescopes, motorized focusers,
telescope accessories, movable mounts
for heavy telescopes.)

Meade Instruments Corporation,
6001 Oak Canyon, Irvine,
CA 92620, U.S.A.
Web: http://www. meade.com.
(Telescopes and accessories of all
kinds.)

Obsession Telescopes,
P.O. Box 804W, Lake Mills,
WI 53551, U.S.A.
Web: http://www.
obsessiontelescopes.com.
(Very large portable telescopes.)

SBIG
(Santa Barbara Instrument Group),
1482 East Valley Road #33,
Santa Barbara,
CA 93150, U.S.A.
Web: http://www.sbig.com.
(Wide range of well-engineered CCD
cameras for amateur and professional
astronomy.)

Starlight Xpress,
Foxley Green Farm, Ascot Road,
Holyport, Maidenhead,
Berkshire SL6 3LA, England.
Web: http://www.starlight xpress.co.uk
(Innovative and cost-effective CCD
cameras for the amateur.)

Tele Vue Optics, Inc.,
100 Route 59, Suffern, NY 10901, U.S.A.
Web: http://www.televue.com.
(Eyepieces and compact refractors of
high quality.)

Vixen Optical Industries Ltd.,
247 Hongo, Tokorozawa,
Saitama 359, Japan.
Web: http://www.vixen.co.jp.
(Telescopes, eyepieces, and
accessories, imported under the Perl,
Celestron, and Orion brand names in
some countries.)

Dealers

USA/Canada

Anacortes Telescopes, 9973 Padilla
Heights Road, Anacortes,
WA 98221, U.S.A.
Web: http://www. buytelescopes.com.

Astronomics,
2401 Tee Circle, Norman,
OK 73069, U.S.A.
Web: http://www. astronomics.com.
(Wide selection of telescopes and
accessories; reliable service. Does not
export.)

Lumicon,
2111 Research Drive #5A,
Livermore, CA 94550, U.S.A.
Web: http://www.lumicon.com.
(Equipment and materials for
astrophotography, including nebula
filters, hypered film, and hypering kits.)

Orion Telescopes and Binoculars,
P.O. Box 1815, Santa Cruz,
CA 95061 U.S.A.
Web: http://www. telescope.com.
(Informative catalogue and web site
with advice for beginners; full line of
telescopes and accessories, including
some Vixen products imported under
the Orion name. Not to be confused
with Orion in Britain, which also
imports Vixen.)

Texas Nautical Repair Company,
3110 S. Shepherd, Houston,
TX 77098, U.S.A.
Web: http://www. lsstnr.com.
(U.S. distributor of Takahashi premium-
quality telescopes and eyepieces.)

University Optics,
P.O. Box 1205, Ann Arbor,
MI 48106, U.S.A.
Web: http://www.universityoptics.com.
(Eyepieces, accessories, and telescope-
making supplies.)

Britain

Telescope House
(Broadhurst, Clarkson & Fuller),
Farringdon Road,
London EC1 M3JB, England.
Web: http://www.telescopehouse.co.uk
(Telescopes, accessories, and books.
Sells its own line of telescopes and
imports products of Meade, Celestron,
Lumicon, and others. Large showroom.)

Orion Optics,
Unit 21, 3rd Avenue, Crewe,
Cheshire CW1 6XU.
Web: http://www. orionoptics.co.uk
(British distributor for Vixen and other
products. Not to be confused with
Orion Telescopes in California, which
also imports Vixen.)

SCS Astro Ltd.,
1 Tone Hill, Wellington,
Somerset TA21 0AU,
England.
Web: http://www.scsastro.co.uk
(Full-line dealer.)

Illustration credits

Photographs

top margin of pp. 8 to 107 :
  photo C. Lehénaff © Archives Larbor
page. 8 and 9 (2 photos) : photo © Archives
  Larbor
page. 10 and 12-13 : photo C. Lehénaff
  © Archives Larbor
page. 17 : photo © Maison de l'Astronomie
page. 19 : photo © Meade Instruments Corp.
page. 23 : photo © App'Ar Studio/Médas-
  Vichy
page 25 (2 photos) and 29 (2 photos) :
  photo C. Lehénaff © Archives Larbor
page 31 : photo H. Burillier © Archives Larbor
page 36-37, 38, 39, 40 and 41 :
  photo C. Lehénaff © Archives Larbor
page 42 : photo © Meade Instruments Corp.
page 43 : photo C. Lehénaff © Archives
  Larbor
page 44 : photo © C. Blanchard
page 45 and 46 : photo © S.pageJ.P. Paralux
page 47 : photo C. Lehénaff © Archives
  Larbor
page 48 : photo © App'Ar Studio/Médas-
  Vichy
page 49 : photo C. Lehénaff © Archives
  Larbor
page 50 : photo © Maison de l'Astronomie
page 51 : photo C. Lehénaff © Archives
  Larbor
page 52 : photo © S.P.J.P. Paralux
page 53 : photo © Meade Instruments Corp.
page 54 : photo © App'Ar Studio/Médas-
  Vichy
page 55 and 56 : photo C. Lehénaff
  © Archives Larbor
page 57 : photo © Optique Unterlinden
page 58 : photo © Meade Instruments Corp.
page 59 : photo © Optique Unterlinden
page 60 : photo © C. Arsidi/Ciel and Espace
page 61 : photo C. Lehénaff © Archives
  Larbor
page 62 : photo © C. Ichkanian/Ciel and
  Espace
page 63 : photo © App'Ar Studio/Médas-
  Vichy
page 64 : photo © Maison de l'Astronomie
page 65 : photo © P. Pellandier/Médas-Vichy
page 66 : photo © Meade Instruments Corp.
page 67 : photo © P. Pellandier/Médas-Vichy

page 69 : photo © C. Ichkanian/Ciel and
  Espace
page 70 : photo © App'Ar Studio/Médas-
  Vichy
page 71 : photo © Maison de l'Astronomie
page 72 : photo © Meade Instruments Corp.
page 73 : photo H. Burillier and P. Chesneau
  © Archives Larbor
page 74, 75, 76-77, 79, 81, 82 (2 photos), 85,
  86 and 87 top : photo C. Lehénaff
  © Archives Larbor
page 87 bottom : photo H. Burillier ©
  Archives Larbor
page 88 and 89 : photo C. Lehénaff
  © Archives Larbor
page 91 left : photo © Maison de
  l'Astronomie
page 91 right : photo C. Lehénaff © Archives
  Larbor
page 92 and 93 : photo © Maison de
  l'Astronomie
page 95 and 96 left : photo C. Lehénaff
  © Archives Larbor
page 96 right : photo © Meade Instruments
  Corp.
page 97, 98, 99, 100 and 101 : photo
  C. Lehénaff © Archives Larbor
page 102 : photo © C. Buil/Ciel and Espace
page 103 : photo © Bayle/Médas-Vichy
page 104 : photo J. Bottand © Archives
  Larbor
page 105 top : photo C. Lehénaff © Archives
  Larbor
page 105 bottom, 106 (2 ph) and 107 :
  photo J. Bottand © Archives Larbor
page 108-109, 110 and 111 :
  photo C. Lehénaff © Archives Larbor

Drawings
page 14, 15, 16, 18, 20, 21, 22, 24, 27, 32, 33,
  35 (2), 68 and 80 : Laurent Blondel